T0073563

PARTICULATE AND GRANULAR MAGNETISM

Particulate and Granular Magnetism

Nanoparticles and Thin Films

K. O'Grady

School of Physics, Engineering and Technology, University of York
Liquids Research Ltd, Mentec, Deiniol Road, Bangor, Gwynedd, LL57 2UP

G. Vallejo Fernández

School of Physics, Engineering and Technology, University of York, York, YO10 5DD,UK

A. Hirohata

School of Physics, Engineering and Technology, University of York, York, YO10 5DD, UK
Center for Science and Innovation in Spintronics, Tohoku University, 2-1-1 Katahira, Sendai 980-8577, Japan

OXFORD
UNIVERSITY PRESS

OXFORD
UNIVERSITY PRESS

Great Clarendon Street, Oxford, OX2 6DP,
United Kingdom

Oxford University Press is a department of the University of Oxford.
It furthers the University's objective of excellence in research, scholarship,
and education by publishing worldwide. Oxford is a registered trade mark of
Oxford University Press in the UK and in certain other countries

Published in the United States of America by Oxford University Press
198 Madison Avenue, New York, NY 10016, United States of America

British Library Cataloguing in Publication Data

Data available

Library of Congress Control Number: 2023936355

ISBN 9780192873118

DOI: 10.1093/oso/9780192873118.001.0001

Printed and bound by
CPI Group (UK) Ltd, Croydon, CR0 4YY

Preface

This text represents the knowledge accumulated by the first author in over 40 years research into particulate and granular magnetic materials. The text is not intended to be a comprehensive text on magnetism as there are other books that adequately cover the subject and which are routinely cited such as the original work by B. D. Cullity (1972), David Jiles (1997), R. C. O'Handley (2000) and most recently K. Krishnan (2016). The text is aimed primarily at experimental chemists, physicists and perhaps electronic engineers and material scientists interested in this subject. For that reason the text has been written with limited theoretical details which are admirably covered in the texts mentioned above. Only the mathematics essential for the understanding of the text is included. Hence the text should be accessible and informative to final year Undergraduate and Masters students and those undertaking PhD research on magnetic materials. It should also provide a teaching aid for those academics charged with this duty and as a text to be used by industrial scientists working in the areas of high-tech magnetism particularly in the information storage industries and those companies concerned with the production of magnetic nanoparticles primarily for biomedical applications.

To that end the text is divided into two parts. Part One covers the basic physics of magnetism from a relatively low level which includes an explanation of some of the unusual terminology in magnetism such as the idea of poles and flux whose origins are little understood. The complexity of the unit systems in magnetism are also presented.

Thereafter a brief review of the principles of domain theory is presented and of particular importance for particulate and granular materials, thermal activation effects and their correct measurement are discussed in some detail. The topic of exchange bias where an antiferromagnetic material is grown in intimate contact with a ferromagnet is presented in significant detail reviewing old theories and numerical models but then focusing on what has become known as the York Model of Exchange Bias which is now universally accepted as the model which describes the behaviour of exchange bias systems when grown in the form of granular thin films.

In Part Two a detailed description of ferrofluids is presented including a simple method for their preparation and the various engineering applications in vacuum seals, loudspeakers, sink float separation and the alignment of non-magnetic entities. A description is provided of the phenomenon of magnetic hyperthermia which is a developing technology with significant potential applications in medical therapies. Other applications of magnetic nanoparticles in biomedicine are also presented.

An extensive discussion of magnetic information storage in conventional recording systems is described including the brief history of the development of this technology whose scale is now enormous as most of the cloud computing system in current use is based on hard drive technology. A forward look to the emerging technology of heat assisted magnetic recording is also given. As magnetic information storage is now beginning a transition to solid state devices, a perspective on Magnetic Random Access Memory (MRAM) is also presented including technology currently transitioning into production in some of the world's leading semiconductor companies with a brief forward look to emerging MRAM technologies.

It is interesting to note that all these technologies depend either on the production of nanoparticles with diameters of around 10 nanometres or the utilisation of thin films often in complex multilayer structures, with perhaps as many as 10 layers in certain devices and storage media. This occurs because it has long been recognised that such engineering specifications cannot be realised in continuous materials due to the inability to control the properties of such materials in thin film form. Hence the focus on particulate and granular materials in the form of nanoparticles and thin films.

Acknowledgements

As indicated in the preface, this text summarises the accumulated knowledge acquired by the first author (KOG) in over 40 years of the study of magnetism. Hence I am happy to acknowledge the opportunity given to me by Dr John Popplewell and the late Dr S. W. Charles formerly of Bangor University in North Wales, who first gave me the opportunity to study magnetism and magnetic materials. I must also acknowledge the significant contribution made to my knowledge and understanding of this subject by Professor R. (Roy) W. Chantrell and the endorsement and encouragement given to me by the late Professor E. P. (Peter) Wohlfarth. Of course much of the work described was undertaken by over 40 PhD and MSc students but the contribution of some was really most significant. In particular Professor Mohammed el Hilo now of Bahrain University must be mentioned. Also my co-author (GVF) and his friend Dr Luis Fernandez-Outon who undertook the basic work on exchange bias described in Chapter 4 warrant particular mention. The work of all the other students is acknowledged by citation of their publications.

I must also acknowledge the financial contributions in terms of grants and facilities made by principally the UK Engineering and Physical Science Research Council and the Commission of the European Union who funded much of the work included in this text. However a number of industrial companies also supported my work and I must mention the former IBM Storage Systems Division, Seagate Media Research in Fremont, Seagate Technology in Northern Ireland and Western Digital Corporation in Fremont, California together with Oxford Instruments here in the UK. The second co-author GVF acknowledges financial support from the Royal Society. AH acknowledges ongoing and significant support from the Japan Society for the Promotion of Science (JSPS) and the Japan Science and Technology Agency (JST). We also acknowledge the financial contribution of Liquids Research Ltd. with whom I am associated, in the production of this text.

A number of individuals have kindly read and commented on various parts of the text or contributed in other ways, particularly the drawing of diagrams by Dr. Will Frost of the University of York. The text has been partially read and commented upon by Dr. Dieter Weller formerly of Western Digital, Ms. Alex Carter of the University of York and my former student Dr. Rob Carpenter now at IMEC in Belgium. My colleagues Prof. Aires Ferreira and Dr. Vijay Patel have provided valuable comments on parts of the text. Finally I must acknowledge the staff and editors of Oxford University Press without whose kind invitation and subsequent input, this book could not have been generated. The image on the front cover was provided by Mr. Ian Helliwell of the University of York.

Contents

Part I

Basic Concepts

Part I

Basic Concepts

1
Concepts, Terminology and Units

The knowledge and study of magnetism goes back more than 2000 years and, as a scientific endeavour, the length of study is possibly only exceeded by that of astronomy. Ancient mariners knew that certain stones when either floated on a liquid or suspended in air would align so that one axis of the piece of material would point towards the North Pole. Certainly in early times no distinction was made between the geographic North Pole and the magnetic North Pole. The best minerals of this type were found in the district of Magnesia in Turkey and the word magnet derives from the Greek name for this region. The mineral itself is now known as magnetite and has been for many hundreds of years.

The first proper scientific study of magnetism was made by William Gilbert (1540-1603) who published his classic text on the magnet (1600) in which he described a picture of the Earth's magnetic field for the first time and the alignment of pieces of magnetite along a north-south axis. Later it was known that touching, or better rubbing, a piece of steel with a piece of the mineral magnetite would also cause that metal to align along the north-south axis when suspended. This remained the case until the linkage between electricity and magnetism was made by Hans Christian Oersted in 1820 who showed that an electric current passed through a piece of wire wrapped around a piece of steel would magnetise it and such a solenoid generated a magnetic field.

Further developments in the study of magnetism and magnetic materials occurred in the late 19th century due to pioneering work by Michael Faraday (1791-1867) and others but the serious study of magnetic materials and an understanding of the phenomena via quantum mechanics was developed in the 20th century.

For this reason many of the concepts of magnetism and in particular the terminology used, are very confusing. Perhaps the best example of this is the use of the term magnetic pole which relates to the alignment of a piece of magnetic material along a south-north axis. The use of the term pole is misleading because what is really meant is magnetic charges where, following the sign convention, we define a pole that points to the north as being positive in nature and that to the south being negative. The term pole is simply used because of the historical concept. However even this is confusing because as every school child knows, a north pole points towards the north. That being the case then the nature of the magnetic charges at the northern extremity of the planet must therefore be negative and hence are a south pole but we all know that in the common vernacular we refer to the northern extremity of Earth as the North Pole. The reason for this comes about because of a misuse of language. The correct description which can be found in many old books on the subject, is the north

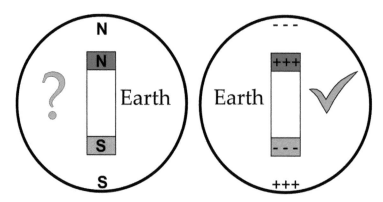

Fig. 1.1: Diagram of the poles of a magnet.

pole of a magnet is correctly described as "*a north seeking pole*". Hence what we call the north pole of a magnet contains positive magnetic charges based on the sign convention as shown in Fig. 1.1. Due to the historical context and various attempts to simplify the use of language, it is no wonder that people find the terminology difficult to understand.

Further confusion when trying to understand magnetic materials arises because the vast majority of work in the field is still undertaken using the cgs system of units. There are a number of reasons for this not least because various industries associated with magnetism and magnetic materials historically worked in the cgs system and when the use of the SI system became almost universal, the industries declined to change. This was partially because the SI system of units comes with its own difficulties. For example if one considered the field from a permanent magnet the SI unit comes out as A/m. This is physically difficult to understand because no current flows in a permanent magnet and it is not clear from where the length term comes. For this reason to this day, most of the work on technological magnetic materials is still described and studies are undertaken using the cgs system. This text will follow this convention but in Section 1.4 we present a table showing the units and conversion factors including the new units recently defined by various standards laboratories and the International Standards Organisation which attempt to relate the units of measurement back to fundamental constants.

1.1 Magnetic Poles, Moments and Fields

As described above the concept of a magnetic pole is confusing when compared with the much more simple concept of a magnetic charge which becomes analogous to an electric charge. None-the-less it is possible to define the force F, between two magnets having pole strengths p_1 and p_2 at a distance d from each other in the cgs system as

$$F = \frac{p_1 p_2}{d^2} \tag{1.1}$$

Equation 1.1 leads to a definition of the pole strength as that which gives rise to a force of 1 dyne when the two poles are positioned 1 cm apart where the dyne is the unit of force in the cgs system. The term pole strength has no units. A magnetic pole produces a field H which exerts a force (F) on a neighbouring pole such that

$$\mathbf{F} = p \cdot \mathbf{H} \text{ and } H = \frac{p}{d^2} \tag{1.2}$$

where d is the distance which defines the unit of field in the cgs system, the Oersted (Oe) such that the force generated on a unit pole is 1 dyne. Michael Faraday defined the concept of field lines whose density gives rise to the force. Hence if a unit pole is surrounded by a sphere of radius 1 cm, the force at the surface is always 1 dyne. The surface area is then $4\pi^2$ and thus $4\pi p$ lines of force emanate from a pole strength p, i.e.

$$1 \text{ Oe} = 1 \text{ line of force/cm}^2 = 1 \text{ Maxwell/cm}^2$$

Of course as established by Maxwell's equations, for all practical purposes there is no such thing as a magnetic monopole unlike the case for electric charges which are of themselves monopoles. The magnetic moment of an object is the fundamental parameter and the energy of a moment $\mathbf{m}$ when exposed to a magnetic field $\mathbf{H}$ is given by

$$E_p = -\mathbf{m} \cdot \mathbf{H} \tag{1.3}$$

In the cgs system of units E_p is in ergs and hence the units of magnetic moment m are ergs/Oersted.

The magnetisation (M) of a material exhibiting a magnetic moment is then defined as the magnetic moment per unit volume (V)

$$\mathbf{M} = \frac{\mathbf{m}}{V} \tag{1.4}$$

This can also be written in terms of the pole strength

$$M = \frac{pl}{V} = \frac{p}{V/l} = \frac{p}{a} \tag{1.5}$$

where a is the cross-sectional area of the magnet. Hence the magnetisation can also be defined as the pole strength per unit of cross-sectional area and has units of ergs/(Oersted·cm^3). However for convenience and abbreviation, this unit is generally written as emu/cm^3 where the emu is the electromagnetic unit of magnetic moment. In the case of nanoparticles and thin films where the magnetic material is never 100% dense and where the actual density is difficult if not impossible to measure, it is more usual to specify the magnetisation per unit mass (σ) in emu/g.

In everyday uses the effects of potentials on other objects is generally described by the word field. The reason why this is confusing is that it is not clear what exactly a field is. The reality is that the problem again is one of language where the correct terminology is "a field of force". From this it is clear that the concepts of a field and its magnitude determines the level of force experienced by another object upon which it acts. In the case of magnetism that is either a current carrying wire or another magnetised or magnetisable object in which the magnetic field induces a moment upon which the force then acts.

1.2 Flux Density and Induction

As is well known there are two sources of a magnetic field deriving either from the flow of a current through a wire or from a material which is magnetised and even in the absence of a magnetising field retains a magnetic moment, which in turn generates a field. Because of the lack of familiarity with field calculations in the cgs system, in this instance equations are presented in both the cgs system and in the SI system. Ironically even those who work in the cgs system often use the SI system when doing field calculations.

A current flowing in a straight wire produces a field where the field lines are circular around the axis of the wire and lie in a plane normal to the axis. The magnitude of this field as a function of distance R from the wire is given by

$$H = \frac{2i}{10R} \text{ (cgs) } ; H = \frac{\mu_0 i}{2\pi R} \text{(SI)} \tag{1.6}$$

where i is the current in Amperes and in the SI system μ_0 is the permeability of free space equal to $4\pi \times 10^{-7}$ Henrys/m. For the case when the wire is curved into a circular loop of radius R the field at the centre along the axis is

$$H = \frac{2\pi i}{10R} \text{ (cgs) } ; H = \frac{\mu_0 i}{2R} \text{(SI)} \tag{1.7}$$

Finally it is possible to produce a small coil which is in effect a helix of wire which produces a much stronger field than a single loop in what is called a solenoid and in this case the field is given by

$$H = \frac{4\pi N i}{10L} \text{ (cgs) } ; H = \frac{\mu_0 N i}{L} \text{(SI)} \tag{1.8}$$

where N is the number of turns and L the length of the solenoid which must be very much greater than its diameter. In the cgs system H will have units of Oersted if R and L have units of cm. It should be noted here that eqn 1.8 applies only to the case where $L >> R$ and to obtain a field in a solenoid which is uniform to any significant extent over say a length of 1 cm requires a solenoid of L/R of 10 or greater.

Such a field is designated as the H-field and is correctly described as a magnetising field. The analogy to the electric case is clear where a simple battery attached to two small plates separated by a distance x generates an electric field given by

$$E = -\frac{dV}{dx} \tag{1.9}$$

Equation 1.9 shows that the gradient of the scalar potential is the E-field.

For the magnetic case considering a simple coil of N turns the scalar potential is $\Sigma N i$ and hence eqn 1.8 or more generally

$$H = -\frac{d\Sigma N i}{dx} \tag{1.10}$$

where dx is an element of the length of the coil. Of course most practical solenoids do not meet the criterion $L >> R$ and the field generated must be calculated using finite element methods such as the well known Finite Element Method Magnetics (FEMM) (https://www.femm.info). An example of such an analysis can be found in Drayton *et al.* (2017).

Following the analogy to the electrical case we now have to look at the value of the total field generated both by the current and for example, by the magnetisation of a material which is inserted into the coil thereby becoming magnetised. The simplest case to think of is that of a solenoid where a coil of wire is wrapped around a pencil, and a current passed through the coil. For a few turns such a coil is quite capable of picking up a paper clip. However if a nail made from mild steel is inserted into the coil by removing the pencil then such an electromagnet, which is what it becomes, would be capable of picking up many paper clips if not a whole box full. Hence the majority of the field is derived from the magnetisation of the steel rod rather than from the flow of current. It is for this reason that conventional electromagnets consist of two large coils with a core of a soft magnetic material such as FeCo which is then capable of generating fields in a gap of a centimetre or so of up to about 30 kOe or 3 T. It is only in the last 50 years or so that superconducting coils with effectively zero resistance, were capable of generating fields of this magnitude and latterly much greater still. Hence the field strength being defined as the density of the force lines resulting in such a system is given by

$$\mathbf{B} = \mathbf{H} + 4\pi\mathbf{M} \ (cgs) \ ; \mathbf{B} = \mu_0 \left(\mathbf{H} + \mathbf{M}\right) \ (SI) \tag{1.11}$$

The symbol B represents the resulting field from either an electromagnet but is often used for the field from a solenoid without a core because that is simply the case when $M=0$. Because this parameter represents the density of the field lines it is often known as the flux density.

It is known to the authors that very few physicists and even those working in the field of magnetism and magnetic materials, are aware of the origin or meaning of the term flux. The correct definition of the term flux in English is a flow of a material but in the case of a permanent magnet nothing is flowing and likewise whilst there is a flow of current in the solenoid, the magnetic field is measured on the axis of the coil and again nothing is flowing. Once again the origin of this term lies in the long history of the study of magnetism. In the 19th and the early part of the 20th century it was believed that electric and magnetic forces derived from the flow of an invisible, tasteless, colourless substance which was defined as the ether. Hence when early scientists tried to describe electrical and magnetic phenomena they talked about the flow of the ether which in the case of magnetism, was generated by either a current flowing in a coil along the axis of the coil or the flow of the ether between two small metal plates which were connected to a source of electric potential. Following the seminal experiment of Michelson and Morley (1887) when it was established that no such liquid or gas existed, they simply dropped the term "of the ether" but continued to talk about the flux of the magnetic field to the confusion of generations of scientists. Reference to old books on electricity and magnetism include a discussion of the ether and how it flows.

When one thinks about this the concept of a permanent magnet with a north and south pole and the flow of the ether from the north pole to the south, it actually gives quite a reasonable description of what is observed. Trying to push two north poles together is like trying to push together two hosepipes with water coming out of them which is difficult and, with a heavy flow of water, impossible. Likewise a north pole is attracted to a south pole because the ether flows out of the north pole and is sucked in by the south pole on another magnet thereby causing the force of attraction. It seems a bit of a shame that such a model was wrong!

With the basic concepts now clear it is easy to understand how a coil with a field flowing through it or the stray field from a permanent magnet results in a force on a body that was previously unmagnetised. Simply the flux density from the coil magnetises a body that has a finite susceptibility thereby inducing a magnetic moment which in turn gives rise to a magnetisation. For the case of a solenoid with a ferromagnetic core the flux lines are correctly defined as lines of induction. This magnetic induction is distinct from the famous Faraday's law of induction where the movement of a magnetised body inside a coil generates an electromotive force (ϵ) in that coil given by

$$\epsilon = -N\frac{d\phi}{dt} \tag{1.12}$$

where N is the number of turns and ϕ is the flux.

However the use of the same term does little to reduce the confusing nature of terminology in magnetism.

1.3 Other Key Concepts

There are a number of other key concepts and parameters which are in some ways analogous to parameters in electrostatics but have unique features because of the dipole nature of magnetic fields as compared with the monopole nature of electrostatic charges.

1.3.1 The Demagnetising Field

Because of the dipole nature of a magnetised body, free poles (or charges) appear usually at the ends of an elongated body as shown in Fig. 1.2. This leads to the circulation of the flux lines from the north pole (positive) to the south pole (negative) thus producing a field outside the magnetised entity which is in the opposite direction to that of the magnetisation within as shown in the figure. This field which is inevitable for all magnetised bodies, is known as the demagnetising field H_d and clearly depends on the shape and the dimensions of the object under consideration. H_d is given by

$$\mathbf{H_d} = -N_d\mathbf{M} \tag{1.13}$$

where N_d is called the demagnetising factor. The two easiest objects to consider for which analytical solutions allow for the calculation of N_d, are that of a prolate spheroid and an oblate spheroid, shown schematically in Figs. 1.3(a) and (b), respectively. As shown in Fig. 1.3, $2a$, $2b$ and $2c$ are the lengths of the three orthogonal axes of the ellipsoid. For each case the exact demagnetising factor along the major axis of the spheroid can be calculated from eqns 1.14 and 1.16 below where $r = c/a$ is the axial

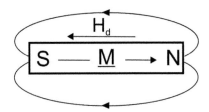

Fig. 1.2: Demagnetising field in a magnetised body.

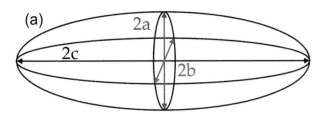

Prolate Spheroid

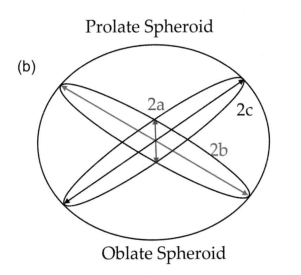

Oblate Spheroid

Fig. 1.3: Schematic diagram of (a) a prolate spheroid and (b) and oblate spheroid.

ratio of the bodies. These formulae apply in the cgs system of units such that the demagnetising factors along each of the three axes a, b and c add up to 4π. In the SI system of units the demagnetising factors add up to 1 and hence a conversion from one system to the other is relatively simple. For the prolate spheroid

$$N_d = \frac{4\pi}{r^2 - 1} \left[\frac{r}{\sqrt{r^2 - 1}} ln \left(r + \sqrt{r^2 - 1} \right) - 1 \right] \tag{1.14}$$

and

$$N_a = N_b = \frac{4\pi - N_c}{2} \quad (a = b < c) \tag{1.15}$$

and for the oblate spheroid

$$N_d = \frac{2\pi}{r^2 - 1} \left[\frac{r^2}{\sqrt{r^2 - 1}} sin^{-1} \left(\frac{\sqrt{r^2 - 1}}{r} \right) - 1 \right] \tag{1.16}$$

and

$$N_a = 4\pi - 2N_c \quad (a < b = c) \tag{1.17}$$

Equation 1.16 generally applies to the case of barium and strontium ferrite particles which when grown chemically form hexagonal platelets with N_a given by eqn 1.17. The demagnetising factor for an oblate spheroid can also be of importance because with modern magnetic measurements instrumentation such as a vibrating sample magnetometer, the sample holder used for samples such as powders is very often in the shape of a small disc consisting of a vessel and a lid and will allow for the correction for the sample shape demagnetising factor. Similarly other instruments use samples in the shape of cylinders so the formula for a prolate spheroid is also of importance.

However in practice it is almost always impossible to produce samples having such perfect shapes. For all practical purposes when considering nanoparticles, samples are generally produced in the shape of a cylinder perhaps of plastic or a machinable ceramic, and the ideal situation occurs when such a cylinder has an axial ratio of >5:1 where $N_c/4\pi = 0.056$ and the demagnetising field can be ignored. Similarly for a magnetic thin films the demagnetising factor for a magnetic measurement in the plane of the film is zero. In these cases the true field acting on the sample when being measured in some form of magnetometer is actually the measured field. However when these criteria are not met then the true field $\mathbf{H_t}$ is

$$\mathbf{H_t} = \mathbf{H_a} - \mathbf{H_d} \tag{1.18}$$

where $\mathbf{H_a}$ is the applied field. Hence a correction must be applied to the field axis to obtain the true magnetisation curve. At two unique points on a hysteresis loop the applied field is the true field and these are the points of the coercivity where $M = 0$. Hence if the coercivity is the only parameter to be measured no correction needs to be applied. However in Fig. 1.4 not only is shown the hysteresis loop but also the line corresponding to $\mathbf{H_d} = -N_d \cdot \mathbf{M}$. Thus it can be seen that in effect, a rotation must be applied to the hysteresis loop and for all materials except those having a perfectly square hysteresis loop, the remanence and the entire switching region of the loop is incorrect unless $H_d \simeq 0$. Figure 1.4 also shows the definition of other key parameters on a hysteresis loop. Tables of selected values of N_d are given in Appendix A.

There is a further unique case which is of particular importance for information storage systems and that is the case where a thin film is magnetised perpendicular to the plane. Due to the demagnetising field, such a system must be composed of a material having extremely high (usually) magnetocrystalline anisotropy (see Section 2.1) along a crystallographic axis aligned perpendicular to the plane of the film. If this

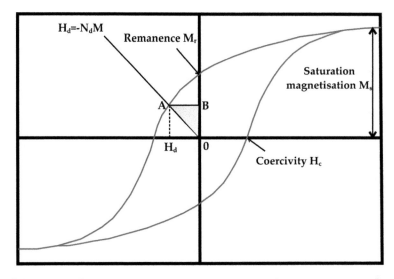

Fig. 1.4: Schematic of a hysteresis loop showing some of the main characteristics and the demagnetising field.

is not the case then such a film will automatically be demagnetised due to the presence of the demagnetising field. However in the last two decades materials meeting this requirement for anisotropy have been developed because it provides significant technological advantages for information storage as will be discussed in Chapter 7.

For such films the demagnetising factors in the plane of the film are zero due to the axial ratio and hence the demagnetising factor perpendicular to the plane is equal to 4π in the cgs system. To put this in context this implies that for a thin film consisting of pure iron which has a magnetisation of 1700 emu/cc, this leads to a demagnetising field which is equal to $-4\pi M_s$ which must be overcome to saturate the sample. This gives a required saturating field of over 21 kOe or in the SI system over 2.1 T. In the plane of the film due to the low magnetocrystalline anisotropy of pure iron, such a film would be expected to saturate in a field of <500 Oe or 0.05 T.

It should be emphasised here that when a sample of a thin film or single crystal is measured in an open magnetic circuit system such as vibrating sample magnetometer (VSM) or a SQUID system, its shape should be a circular disc. If an angular shape such as a square or any other with sharp corners is used, the demagnetising field will be highly non-uniform and particularly large near the corners. Under these conditions H_d can prematurely nucleate a reverse domain and if a granular film has exchange coupled grains as discussed in Section 3.6, or the sample is a single crystal, this can give rise to false values for parameters such as the coercivity and remanence.

This tendency for a magnetised body to demagnetise itself is of critical importance in information storage because each bit of information consists of an ultra-small magnet. The presence of not only the demagnetising field from the bit, but also the demagnetising field from neighbouring bits, leads to all bits of stored information experiencing a strong and often complex demagnetising field. This can lead to a thermal decay of the magnetisation and hence of the strength of the field emitted by the bit. In the terminology of the sector this is known as thermal loss of data. Thermal effects are discussed in detail in Chapter 3. It is for this reason that materials used in information storage are generally engineered to have low values of M_s.

1.3.2 Susceptibility and Permeability

Other key parameters which are of particular relevance for the study of fine particles are the susceptibility (χ) and permeability (μ_r) of a material. The susceptibility of a material is defined as the slope of a magnetisation curve (M vs H) or hysteresis loop at each point around the loop. Hence the susceptibility is field dependent. It is often the case that people refer to the susceptibility as being that value close to the origin (χ_i) known as the initial susceptibility, which for certain materials is a fixed value where the coercivity and remanence is zero. However this is not the case for a material exhibiting magnetic hysteresis. None-the-less susceptibility is simply defined as

$$\chi = \frac{dM}{dH} \tag{1.19}$$

For materials not exhibiting hysteresis which exhibit quasi-paramagnetic behaviour the variation of the initial susceptibility χ_i with temperature is of particular importance because it can be used to indicate the effect of thermal energy on the magnetic moments in e.g. a sample of ultrafine particles.

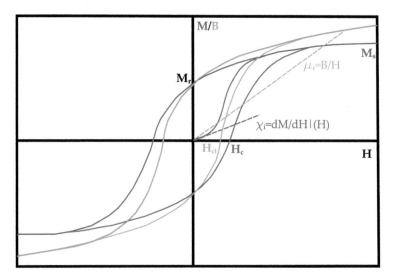

Fig. 1.5: Schematic of a B-H and M-H curves.

However for engineers concerned with the development of technology, a parameter of greater significance is the total flux density B which as we have seen in eqn 1.11 results from the applied field and the magnetisation of the sample. Shown in Fig. 1.5 is a hysteresis loop in the form of a M vs H curve showing characteristic parameters. Also shown in Fig. 1.5 is the total flux density B vs H derived from the magnetisation and the magnetising field H. The first clear difference is that the B-H loop does not saturate. Also it is clear that the value of remanence obtained is the same because at this point the H field is zero. However the coercivity is now lower due to the presence of the applied field. The coercivity of an M-H loop is correctly called the intrinsic coercivity.

The slope of such a B-H curve is known as the permeability (μ_r) and from eqn 1.11 in the cgs system of units the two parameters are related by $B/H = 1 + 4\pi (M/H)$ hence $\mu_r = 1 + 4\pi\chi$. It should be noted that in this definition the parameter μ_r is not the slope $\mathrm{d}B/\mathrm{d}H$ of the B-H curve but is the slope of a line from the origin to a particular point on the curve, i.e. $\mu_r = B/H$.

1.4   Units and Conversion Factors

Introduced in 1960, the international System of Units (SI) was the successor to Giorgi's 1901 rationalised metre-kilogram-second system. It is the preferred system of units across many Physics disciplines with the main exception to this being magnetics, where the centimetre-gram-second (cgs) system has been used for well over a century. Conversion between SI and cgs is generally straightforward except in the case of electromagnetic phenomena. A revised SI system was introduced in 2019 in which the permeability of vacuum, μ_0, is no longer a universal constant. Instead, Planck's constant (h) and the charge of the electron (e) have fixed values. While μ_0 is equal to 1 in the cgs system, it is a measurable parameter in the revised SI system of units. For a more detailed description on the history of magnetic units and the pros and cons of each system see Goldfarb (2017 and 2018). A guide to convert several commonly used magnetic quantities from one set of units to another is given in Table 1.1. The $\hat{=}$ symbol in the 'Conversion Factor' column highlights the different dimensions of the cgs and SI systems of units (Page, 1970). For instance, it is incorrect to say that 1 T equals 10 kOe. The correct expression is 1 T corresponds to 10 kOe.

Table 1.1 SI and cgs units for different magnetic quantities

Quantity	Symbol	SI	cgs	Conversion Factor
Magnetic flux density, magnetic induction	B	T	G	$1\,\mathrm{T} \cong 10^{-4}\,\mathrm{G}$
Magnetic flux	ϕ	V·s	Mx	$1\,\mathrm{V\cdot s} \cong 10^{8}\,\mathrm{Mx}$
Magnetic field strength, magnetising force	H	A/m	Oe	$1\,\mathrm{A/m} \cong 4\pi\cdot10^{-3}\,\mathrm{Oe}$
Volume magnetisation	M	A/m	emu/cm³	$1\,\mathrm{A/m} \cong 10^{-3}\,\mathrm{emu/cm^3}$
Mass magnetisation	σ	A·m²/kg	emu/g	$1\,\mathrm{A\cdot m^2/kg} \cong 1\,\mathrm{emu/g}$
Magnetic moment	m	A·m²	emu = erg/Oe	$1\,\mathrm{A\cdot m^2} \cong 10^{3}\,\mathrm{emu}$
Volume susceptibility	χ_v	dimensionless	dimensionless	$\chi_v\ (\mathrm{SI}) \cong \chi_v\ (\mathrm{cgs})/4\pi$
Mass susceptibility	χ_m	m³/kg	cm³/g	$1\,\mathrm{m^3/kg} \cong 1/(4\pi\cdot10^{-3})\,\mathrm{cm^3/g}$
Molar susceptibility	χ_m	m³/mol	cm³/mol	$1\,\mathrm{m^3/mol} \cong 1/(4\pi\cdot10^{-6})\,\mathrm{cm^3/mol}$
Permeability	μ	N/A²	dimensionless	$1\,\mathrm{N/A^2} \cong 1/(4\pi\cdot10^{-7})$ dimensionless
Relative permeability	μ_r	dimensionless	not defined	
Permeability of free space	μ_0	N/A²	dimensionless	$4\pi\cdot10^{-7}\mathrm{N/A^2}$ (SI), 1 (cgs)
Energy product	W	J/m³	erg/cm³	$1\,\mathrm{J/m^3} \cong 10\,\mathrm{erg/cm^3}$
Demagnetising factor	N_d	dimensionless	dimensionless	1 (SI), 4π (cgs)
Bohr magneton	μ_B	0.927×10^{-23} A·m²	0.927×10^{-21} erg/G	

2
Magnetic Domains

The hypothesis of the existence of domains was proposed by Pierre Weiss in 1906 but little work of significance was undertaken on his theory until 1949 when Williams *et al.* at Bell Labs showed that the theory was in essence correct. From that time domains and domain wall motion became the accepted model of magnetisation reversal in bulk materials. The best modern text on domains is the seminal work *Magnetic Domains* by Hubert and Schäfer (1998).

The principle of why magnetic domains form is clear and derives from the demagnetising field discussed in Section 1.3.1. Basically the presence of a demagnetising field on a magnetised body, as shown in Fig. 1.2, generates a magnetostatic energy. This energy is given by the area of the triangle OAB in Fig. 1.4 which is

$$E_{ms} = -\frac{1}{2}\mathbf{H_d}\mathbf{M}(cgs) \qquad (2.1)$$

Given that $\mathbf{H_d} = -N_d\mathbf{M}$

$$E_{ms} = \frac{1}{2}N_dM^2 \qquad (2.2)$$

This magnetostatic energy can be reduced substantially by the formation of different regions of a ferro- or ferri-magnet breaking up into domains. Within the domains the individual magnetic moments on atoms or grains are all aligned in one direction but in neighbouring regions or domains, the direction of the overall magnetisation is different. In the ideal case the magnetic domains will form a closed magnetic circuit known as flux closure, such that there is no external demagnetising field although this in itself will depend upon the shape of the sample. For example if the sample of a thin film or single crystal is square or rectangular, it will be very difficult to achieve flux closure in the right-angle corners of the material. None-the-less such flux closure configurations have been observed in very carefully prepared samples. Figure 2.1 shows a schematic of such a flux closure configuration which has been observed in a single crystal of iron with the domains along (100) and (010) directions.

The direction of the magnetisation within domains is determined by a phenomena known as the anisotropy of a magnetic material. The term anisotropy simply means that the properties of a ferro- or ferri-magnetic material vary along different directions within the material. These directions can be caused by the nature of the crystal structure of the material where the distance between atoms or the bond strength are different, leading to different properties due to the variation in the exchange coupling between the atoms.

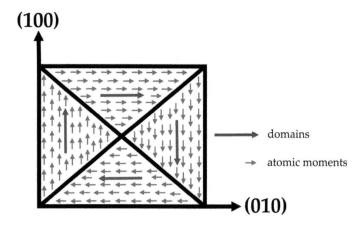

Fig. 2.1: Schematic representation of closure domains in an iron single crystal.

Similarly for the case of isolated magnetic nanoparticles which generally contain a single magnetic domain as discussed in Section 2.4, the shape of the particle itself can lead to a variation in the anisotropy due to the variation of the effective demagnetising field particularly where the particles are e.g. an oxide, such that no intergranular exchange coupling can occur (see Section 2.7). In such particles it is generally the case that the shape anisotropy produces a preferential direction for the moment along the long axis of the particles. Of course there are exceptions where a nanoparticle has a very high magnetocrystalline anisotropy that overcomes the shape anisotropy effect. A good example of this type of material is barium or strontium ferrite which have a hexagonal structure and where the particles tend to grow as hexagonal platelets with the preferred direction for the moment perpendicular to the plane of the platelet. Hence for all cases of magnetic domains in particulate and granular materials, it is the nature and magnitude of the anisotropies that must be considered.

2.1 Magnetocrystalline and Shape Anisotropy

It is not the intention here to provide a comprehensive description of all magnetic anisotropies some of which do not apply to particulate and granular materials. For the purposes of understanding the behaviour of such materials it is almost always the case that only magnetocrystalline and shape anisotropy are relevant. It is also not the intention to provide a detailed mathematical analysis of anisotropies because many of the original theories which date back many decades, are based upon the concept of the anisotropy being single valued and uniform. In the case of particulate and granular materials that is rarely the case because of effects such as a particle or grain size and shape distributions discussed in detail in Section 3.3. It is also generally the case that the direction of the preferred magnetic axis, known as the easy axis, is not the same for all particles and granules. This also leads to a distribution of the effective anisotropy in the sample of material to be measured. A further complication occurs in that many of these early theories of domains and the role of anisotropy take no account of the effect of thermal activation and are in general theories that apply at zero Kelvin only. Hence many of the experimental techniques described which are based on measurements taken around room temperature, are not necessarily applicable to particulate and granular materials.

For those interested in the fundamental behaviour and origins of anisotropy there are many suitable texts such as those due to Chikazumi (1997), Cullity and Graham (2008), Jiles (1991) and O'Handley (1999) to which reference should be made.

2.1.1 Magnetocrystalline Anisotropy

The physical origin of magnetocrystalline anisotropy rests in the coupling of the angular (L) and spin (S) moments of the atoms in the crystal structure. The crystal structure of the material dictates the direction of the chemical bonds between the atoms which in turn orients the orbital angular momentum (L). The magnetic moment of the atoms derives from the spin on the electrons and when the angular momentum becomes oriented, likewise the spin angular momentum becomes oriented due to L-S coupling. When a magnetic field is applied, the spin angular momentum wishes to reorient to align with the magnetic field and this reorientation of the spins is opposed by the L-S coupling in the orbitals. Because the bonding between neighbouring atoms is not uniform in different directions, this gives rise to anisotropic behaviour.

The best example of this is to consider a circular sample in the form of a disc of bcc iron cut out from the plane formed by [001] and [110] directions. Such a sample will contain directions within it along the [001], [110], and [111] directions as shown in Fig. 2.2(a) which also shows the location of the atoms in the bcc structure of Fe. Magnetic measurements can now be made along each of these directions which are shown schematically in Fig. 2.2(b) (Cullity and Graham, 2008). The original measurements were made by Honda and Kaya in 1926.

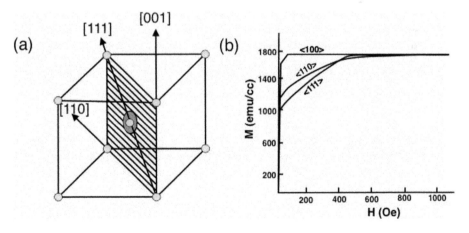

Fig. 2.2: (a) Three principal crystallographic directions in the (110) of an Fe crystal and (b) corresponding hysteresis loops measured along each of those directions (Honda and Kaya, 1926).

As can be seen from the magnetic measurements, bcc iron can most easily be magnetised along [100] directions. The behaviour in the other directions is somewhat more complicated where the [110] direction initially magnetises very easily as compared with the [111] directions but at high fields above 400 Oe, it is easier to saturate along the [111] direction. Because of this behaviour [100] type directions i.e. along the cube edges, are defined as the easy directions with [110] being defined as medium and [111] being defined as a hard direction due to the behaviour in low fields.

Because the rotation of the magnetic moment away from an easy direction requires work to be done, this gives rise to the magnetocrystalline anisotropy energy E_K. This energy is generally written as an energy density, i.e. the energy per unit volume of the crystal (E_K/V), and for magnetoscrystalline anisotropy in cubic crystals can be expressed as a series expansion of the direction cosines of M_s relative to the crystal axis. The parameter K is known as the anisotropy constant with units of ergs/cc (cgs) or J/m^3 (SI). Assuming M_s makes angles a, b, c with the crystal axes and the direction cosines are given by α_1, α_2 and α_3 then

$$E_K/V = K = K_0 + K_1 \left(\alpha_1^2\alpha_2^2 + \alpha_2^2\alpha_3^2 + \alpha_3^2\alpha_1^2\right) + K_2 \left(\alpha_1^2\alpha_2^2\alpha_3^2\right) + ... \qquad (2.3)$$

In general the term K_0 is a constant in all directions and can generally be ignored. Due to the fact that the direction cosines are generally less than 1 and raised to the power 2 before being multiplied together, higher order terms are almost always ignored. For bcc iron the value of K_1 is 4.8×10^5 ergs/cc and $K_2 = \pm(0.5\times10^5)$ ergs/cc. Similar studies for single crystals of nickel which has an fcc structure, find that the easy direction is along [111] with the medium direction being along [110] and a hard direction along the cube edges [100]. For this case the value of K_1 becomes negative.

The situation for other crystal structures can be quite complex but in the case of particulate and granular materials these are generally cubic or exhibit an hexagonal close packed (hcp) structure. Good examples of such materials that are found in particulate and granular form would be metallic iron, nickel and fcc and hcp cobalt and their alloys, and ferrimagnets including barium or strontium ferrite. Cobalt has a high temperature phase which is fcc and the cubic phase is commonly found in fine particles or thin films particularly where the cobalt is in contact with or contaminated by, another cubic material. This situation arises because hcp and fcc structures are both close packed configurations consisting of 72% solids and hence the energy difference between the two structures is relatively small. It is also the case that with thin films containing cobalt, which are generally grown on seed layers which may themselves be cubic, are subject to stacking faults for the reason given above and as we will see in Chapter 7, great care has to be taken with the structure of thin films based on cobalt alloys, which form the basis of the recording material in hard disk drives, to ensure that stacking faults do not occur.

Figure 2.3(a) shows the crystal structure of hcp cobalt and Fig. 2.3(b) the resulting magnetisation curves along the easy c-axis and any hard axis lying in the basal plane. In the case of hcp cobalt the anisotropy is very high at 4.5×10^6 ergs/cc meaning that the energy difference between the moments being aligned along the long axis and any other axis is very large. A similar situation exists for barium ferrite. Because there are

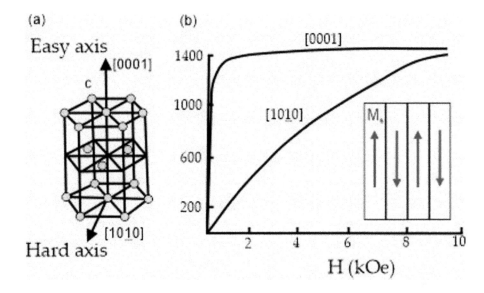

Fig. 2.3: (a) Schematic crystal structure for cobalt and (b) magnetisation curves for single crystal cobalt along the easy and hard axes.

now only two directions the variation of the anisotropy with angle is greatly simplified and of the form

$$E_K/V = K_0 + K_1 cos^2\theta + K_2 cos^4\theta \qquad (2.4)$$

This means that domains all lie along the c-axis and flux closure is not achieved as shown schematically in the inset of Fig 2.3(b).

It should be noted that magnetocrystalline anisotropy is temperature dependent. This is important because it allows for thermally assisted switching of moments as in the emerging technology of Heat Assisted Magnetic Recording (HAMR), discussed in Section 7.6.

The temperature dependence of magnetocrystalline anisotropy has complex origins and there is poor agreement between theory and experiment. However it is generally the case that K_1 decreases with increasing temperature assuming that no phase change occurs in the material. Those with an interest in this topic are referred to the text on *The Physics of Magnetism* by Chikazumi (1997). At elevated temperatures and due to the thermal energy in the system, the local magnetic moments can point in a direction slightly off the easy axis. In general there will be a distribution of directions. This increases the total energy along the easy axis direction and decreases it along the hard axis leading to a temperature dependence of the magnetocrystalline anisotropy of the form

$$\left[\frac{K_1(T)}{K(0K)}\right]^n \propto \left[\frac{M_s(T)}{M_s(0K)}\right]^{\frac{n(n+1)}{2}} \qquad (2.5)$$

where n is the power of the anisotropy function. In the case of uniaxial anisotropy $n = 2$ so

$$\frac{K_1(T)}{K_1(0K)} \propto \left[\frac{M_s(T)}{M_s(0K)} \right]^3 \tag{2.6}$$

while for cubic anisotropy $n = 4$ resulting in

$$\frac{K_1(T)}{K_1(0K)} \propto \left[\frac{M_s(T)}{M_s(0K)} \right]^{10} \tag{2.7}$$

Table 2.1 shows the room temperature values for the magnetocrystalline anisotropy of different magnetic materials.

Table 2.1 Room temperature value of the anisotropy constant(s) for a selection of magnetic materials.

Material	Crystal Structure	K_1 ($\times 10^5$ ergs/cm^3)	K_2 ($\times 10^5$ ergs/cm^3)
Fe	bcc	4.8	± 0.5
Ni	fcc	-0.5	-0.2
Co	fcc	5.4	
$(Ni_{0.5}Zn_{0.5})O \cdot Fe_2O_3$	Spinel	1.2	
$FeO \cdot Fe_2O_3$	Spinel cubic	-1.1	
$MnO \cdot Fe_2O_3$	Spinel cubic	-0.3	
$NiO \cdot Fe_2O_3$	Spinel cubic	-0.62	
$MgO \cdot Fe_2O_3$	Spinel cubic	-0.25	
$CoO \cdot Fe_2O_3$	Spinel cubic	20	
Co	Hexagonal	45	15
$BaO \cdot 6Fe_2O_3$	Hexagonal	33	
YCo_5	Hexagonal	550	
$SrO \cdot 6Fe_2O_3$	Hexagonal	8	
$SmCo_5$	Hexagonal	1100-2000	
$Nd_2Fe_{14}B$	Tetragonal	500	
FePt	Tetragonal	700	
FePt	$L1_0$	500	

2.1.2   Shape Anisotropy

Shape anisotropy arises as a direct consequence of the presence of a demagnetising field as discussed in Section 1.3.1. When considering particulate and granular materials produced either by chemical methods or any method of evaporation or sputtering, we are generally considering particles that exist in a single domain state and are hence at saturation at all times. If the particle is not perfectly spherical this will result in a magnetostatic energy which varies with direction. Typically it is easiest to consider a particle that is deliberately elongated along a given direction or perhaps grows naturally with a non-perfectly spherical shape. This gives rise to a self-demagnetising field discussed in Section 1.3.1 which in turn gives rise to a variation in magnetostatic energy. Such a particle can be approximated to a cylinder which in turn approximates to a reasonable degree with a prolate spheroid. For this case it is possible to use the exact calculation of the demagnetising factor (N_d) along the long axis of such a spheroid using the formula given in eqn 1.14 reproduced here where r is the axial ratio, $r = c/a$.

$$N_c = \frac{4\pi}{r^2 - 1} \left[\frac{r}{\sqrt{r^2 - 1}} ln \left(r + \sqrt{r^2 - 1} \right) - 1 \right] (a = b \neq c) \tag{2.8}$$

and

$$N_a = N_b = \frac{4\pi - N_c}{2} \tag{2.9}$$

where c is the major axis of the spheroid and a and b are the two minor axes orthogonal to c. The demagnetising factors for prolate and oblate ellipsoids for a range of aspect ratios are included in Appendix A.

Of course it is not necessarily the case that in a particle or grain, any elongation in the shape of an individual particle will coincide with the easy direction within the crystallites. Hence there can be instances where there are mixed anisotropies in a material. For those materials having relatively low magnetocrystalline anisotropy such as iron where $K=4.8\times10^5$ ergs/cc or magnetite where $K = 1.1\times10^5$ ergs/cc, shape anisotropy normally dominates the magnetocrystalline term. In all cases the shape anisotropy constant K_s is given by

$$K_s = \frac{1}{2}(N_a - N_c)M_s^2 \tag{2.10}$$

In eqn 2.10 the importance of the term M_s^2 should be noted. For nickel and the ferrites the value of M_s is typically in the range 350 to 480 emu/cc but, due to the low values of K_1, leads to dominant shape anisotropy for particle elongations of around 10%. The exceptions to this rule of thumb are barium and strontium ferrite where the hexagonal structure leads to a large uniaxial magnetocrystalline anisotropy and their low values of M_s make shape anisotropy generally insignificant. There is another exception for the cubic spinel cobalt ferrite ($CoO.Fe_2O_3$). This material has a value of K_1 of 2×10^6 ergs/cc compared with values around or less than 1×10^5 ergs/cc for the other cubic ferrites. This anomaly arises due to the complex distribution of the Co^{2+} and Fe^{3+} ions between the tetrahedral and octahedral sites in the lattice which gives an inherently high anisotropy making the effects of shape anisotropy less significant. The structure of ferrites is discussed in Section 2.7.

2.2　Principles of Domains

We now have the concept of why domains form which has been verified by experiments. We must now consider the structure of the regions between domains commonly known as domain walls and how such a structure allows for magnetic alignment and how it proceeds.

2.2.1 Direct Exchange Interactions

In the interest of clarity for those who are not familiar with quantum physics, the exchange interaction is that which requires the spins on electrons to align either parallel or anti-parallel. If we consider two isolated hydrogen atoms as shown in Fig. 2.4(a) then the spin on the single electron in each atom can point either up or down corresponding to the spin being $\pm 1/2$. However when the two atoms combine to form a hydrogen molecule (H_2) as shown in Fig. 2.4(b), then the Pauli Exclusion Principle requires that one electron has spin up and the other has spin down. Since the molecule has formed and both electrons are in a 1s orbital, each electron is not associated with its parent proton and hence can be *exchanged* between the two protons. It is this form of exchange that gives the title to the exchange interaction. The exchange energy for two electrons first defined by Heisenberg (1928), is given by

$$E_{ex} = -J_{ex}\mathbf{S_1} \cdot \mathbf{S_2} \tag{2.11}$$

where J_{ex} is the exchange integral and $\mathbf{S_1}$ and $\mathbf{S_2}$ are the spin angular momenta of the two electrons.

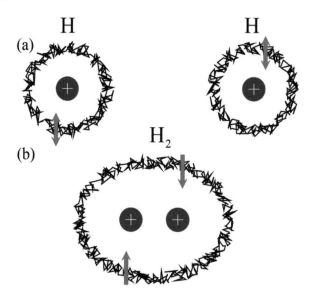

Fig. 2.4: Schematic of (a) two isolated hydrogen atoms and (b) the H_2 molecule.

If J_{ex} is positive then parallel spin alignment is favoured promoting ferromagnetism whereas when it is negative antiferromagnetic order results as is the case for Mn. It should be noted here that the direct exchange interaction is electrostatic in nature coupled with the quantum nature of the energy levels of electrons, i.e. it is not magnetic in origin. Many major figures in theoretical solid state physics contributed to the development of our understanding of exchange interactions. Notably amongst these were Hans Bethe (1933) and John Slater (1930) who showed that under certain assumptions J_{ex} changes from negative to positive and back and identified the three ferromagnetic elements. The Bethe-Slater curve is shown in Fig. 2.5.

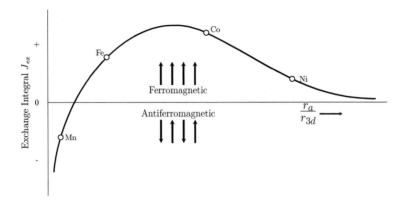

Fig. 2.5: Bethe-Slater curve.

In practice the calculation of values of J_{ex} is highly complex for real crystals consisting of many atoms and is beyond the scope of this text. The interested reader is referred to texts having a greater theoretical emphasis such as that by R. C. O'Handley (2000). Suffice to say that the consequence of the exchange interaction when it promotes the parallel or antiparallel alignment of spins must always lower the overall energy of the system.

2.2.2 Domain Walls

The region between domains consists of a gradual rotation of the spin directions on individual atoms as shown schematically in Fig. 2.6 for a 180° wall. The spin rotation must be gradual to minimise the exchange energy generated by having neighbouring spins not aligned parallel to each other. However those spins not aligned in an easy direction generate anisotropy energy, which would favour an abrupt or narrow domain wall. It is the competition between these two requirements that gives rise to the domain wall width.

There is an important distinction to draw between domain walls in a bulk sample which can have spin rotation in three dimensions and those in thin films whether granular or single crystal. Due to the strong demagnetising field perpendicular to the plane, a 2-D or Néel wall results. A 3-D wall is called a Bloch wall.

2.2.3 Domain Wall Width

For two spins misaligned by an angle θ the exchange energy (E_{ex}) is

$$E_{ex} = -2J_{ex}S^2 cos\theta \tag{2.12}$$

where J_{ex} is the exchange integral and S is the value of each spin. Now the series expansion for $cos\theta$ is

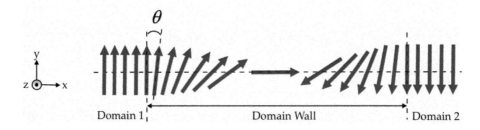

Fig. 2.6: Schematic of a Néel domain wall with the magnetic moments within the wall rotating in the plane of the substrate.

$$cos\theta = 1 - \frac{\theta^2}{2} + \frac{\theta^4}{24} - ...$$ (2.13)

Therefore

$$E_{ex} = J_{ex}S^2\theta^2 - 2JS^2 + ...$$ (2.14)

If the wall is N atoms thick then per unit area there will be $1/r^2$ rows of N atoms. Therefore the energy per unit area is

$$E_{ex} = J_{ex}S^2\theta^2 N\frac{1}{r^2}$$ (2.15)

Some spins will now lie away from the anisotropy easy axis. This gives an anisotropy energy KV which per unit area of the wall is KNr.

$$\therefore E_{tot} = E_{ex} + E_K$$ (2.16)

$$E_{tot} = \frac{J_{ex}S^2\pi^2}{Nr^2} + Kt_w$$ (2.17)

The minimum energy E_{min} is dE_{tot}/dt_w

$$E_{min} = -\frac{J_{ex}S^2\pi^2}{t_w^2} + K = 0$$ (2.18)

where t_w is the width of the domain wall.

$$\therefore t_w = \sqrt{\frac{J_{ex}S^2\pi^2}{Kr}}$$ (2.19)

The minimum wall thickness is when the two energy terms are equal. This means that exchange wants to widen the wall to keep θ small while K wants to narrow it to reduce anisotropy energy.

J_{ex} is proportional to the Curie Temperature T_c and is approximately $0.3kT_c$.

$$\therefore t_w = \sqrt{\frac{0.3kT_c\pi^2}{4K}} \tag{2.20}$$

For bcc iron this gives a wall width of approximately 30 nm or ~120 atoms. Therefore the angle between each spin θ is

$$\therefore \theta = \frac{180}{120} = 1.5° \tag{2.21}$$

For nickel t_w= 72 nm, i.e. ~290 atoms and $\theta = 180/290 \sim 0.62°$ due to the lower anisotropy of Ni. For uniaxial hcp cobalt only 180° walls exist and due to the very high anisotropy t_w= 25 nm or ~8 atoms.

2.2.4   Domain Wall Motion

The behaviour of domain walls in any system is extremely complex as we will see. However there are a number of simple cases that can be considered, principle amongst which is that for a single crystal. As mentioned previously the sample should be cut in a circular manner with well defined crystal planes on a surface to be observed. In such samples it has been observed that if a field is applied in the direction of one domain within the crystal, that domain will grow at the expense of the other domains by a process of domain wall motion.

At all times the sample attempts to maintain a flux closure configuration but none-the-less with a net magnetisation due to the growth of the domain oriented in the field direction. Domain wall motion proceeds via a rippling of the misaligned spins as they translate through the sample. The motion is in a similar manner to that that would be observed in a small wave on a calm ocean moving progressively across its surface. This situation is shown schematically in Fig. 2.7 which is a representation from actual experimental data due to De-Blois (Cullity, 1972). This worker observed the domain pattern represented in the schematic in an Fe-Co alloy platelet that was 165 microns wide. The domain walls were observed by sprinkling a colloid of very small magnetic particles onto the surface which then gathered at the field gradients around the domain walls enabling them to be viewed. Note that this platelet was indeed square so there will have been effects from the strong demagnetising fields in the corners but none-the-less the domain wall motion process is quite clear. However in the experiments reported by De-Blois the sample was not taken to saturation presumably because the small domains would not have been visible at the edge of the sample. Under those circumstances when the domains reach the edge of the sample so that the domain walls cannot propagate further, it is necessary for the spins within those residual domains, which for a square sample would almost certainly be in the corners, to reverse by a rotation mechanism of the spins which in turn would involve overcoming an anisotropy barrier as well as the local demagnetising field.

As long ago as 1919 Barkhausen observed that the motion of domain walls through a magnetised body was not a smooth process but rather consisted of a number of jumps. This observation was made by wrapping a coil around a sample as it was magnetised and using some form of amplifier and loudspeaker when a series of clicks

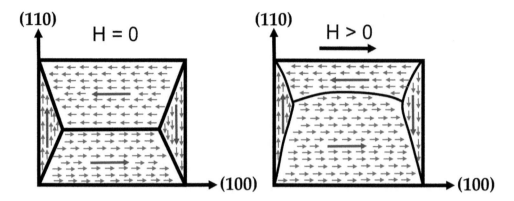

Fig. 2.7: Schematic representation of the domain configuration observed by De-Blois for a Ni-Co platelet (Cullity 1972).

were heard as suddenly there was a change in the flux thereby inducing a voltage in the coil. These days such effects can be observed very clearly by the use of an oscilloscope.

At the time, the origin of these clicks was not well understood but subsequently it was realised that as the domain walls attempted to propagate in the ripple manner described above, any defects within any sample could give rise to the motion of a domain wall being impeded. Such defects are known by the generic term domain wall pins. There are many origins of domain wall pins and only a brief summary of the most important ones can be given here. Furthermore as this text is concerned with particulate and granular magnetism, in general the impediment to domain wall motion occurs at the boundaries of the grains within a granular thin film.

One of the most common defects that occurs in almost all but the best prepared single crystal samples, is the presence of inclusions. These are foreign atoms that are possibly not ferromagnetic but can also include areas where there may be a trace of an oxide or sulphide in the material resulting from the preparation process, but none-the-less these inclusions are generally non-magnetic entities. In essence such entities even in what might appear to be a fully dense single crystal, provide the opportunity for free magnetic poles to occur thereby creating a very localised and very strong demagnetising field within the materials. For example if one considers the possibility of an inclusion of say 5 nm then the proximity of the north and south poles would give rise to a very strong demagnetising field.

In a similar manner any void occurring in the crystal would also give rise to a similar effect which would be likely to pin a domain wall and inhibit its smooth movement. Such voids can come about simply because of a crystal defect where there is a missing atom and the absence of such an atom would then reduce the exchange interaction between neighbouring atoms which promotes the smooth movement of the domain wall. Even dislocations within the crystal where there is a slippage between neighbouring crystal planes would give rise to a similar effect. However in granular materials forming thin films there are inevitably voids and crystallographic discontinuities at the surface of the grains. Also at the surface of the grains there are likely to be traces of oxides or hydroxides produced from the atmosphere or chemical process used to prepare the film. These effects will dominate all others. In a similar way if the film has been or is under any stress this will change the interatomic distances within the crystal and again that separation of distance will have a significant effect upon the exchange interaction which varies approximately as $1/r^6$ where r is the interatomic distance.

As this text is concerned primarily with the behaviour of granular thin films, in addition to grain boundary effects there are also significant domain wall pins that can occur in both granular thin films and even single crystal thin films. These derive from roughness effects which act like an inclusion or a void in the resulting material. Such effects can occur between the substrate upon which the single crystal is grown or due to any level of surface roughness particularly if a capping layer or even a seed layer beneath the magnetic material has been used.

Inevitably because of their random nature and a difficulty of identifying such small defects, the effects are extremely difficult to predict or quantify but none-the-less the behaviour of materials reversing by domain wall motion is extremely complex and of some critical importance in many aspects of modern technology. It is for these reasons that almost all materials used in advanced magnetics technology such as hard drives and the emerging MRAM (see Chapter 8), are produced by sputtering or some other form of ion beam technology. Such films consist of single domain grains around 10 nm in diameter (see Section 2.4). However if such grains are exchange coupled this can lead to domain based behaviour (see Section 2.6). These granular films have properties that can be readily and reproducibly adjusted and deposited rapidly with a high degree of automation and continuous production.

The final issue regarding domain wall motion is the speed at which the domain wall can move through a sample. In addition to the effects of defects and other domain wall pins the speed of a domain wall motion is affected both by the exchange interaction within the material and the value of the anisotropy. Typically domain walls move at a speed of between 1 and 10 cm/sec/Oe. However certain specialised materials, particularly those used as flux multipliers in high frequency inductors, can be carefully engineered and annealed to generate domain wall velocities of up to 3 m/sec/Oe. Such materials are usually very complex alloys of nickel with inclusions of material such as zirconium but with individual particles being grown to micron sizes so that they are multidomain and hence reverse by domain wall motion. These are often coated in an organic binder so that each individual grain does not interact with its neighbours and hence crystallographic uniformity of the material is only required over micron distances. For very high frequency inductors because of eddy current and skin depth effects, which is the case described above, these are suppressed by the inclusion of an insulating binder. An alternative is to use micron size particles of a very soft ferrite material such as Ni-Zn ferrite. However the properties of such materials do not lie within the scope of this text.

2.3 Magnetisation Processes

Now we have a concept of the mechanism by which domains form and behave, Fig. 2.8 shows schematically a portion of the measurement of a hysteresis loop for a material containing domain walls when the sample was perfectly demagnetised at the beginning of the process. Only the first quadrant of the hysteresis loop is shown for clarity.

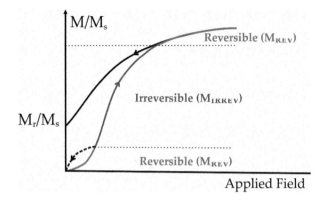

Fig. 2.8: Schematic of the reversible and irreversible parts of the magnetisation.

When an initial generally very small field is applied, those domains oriented in the direction of the applied field will begin to grow, particularly if the sample is a single crystal of quite high quality. If the field is then removed the domain wall will relax back to the initial condition. Hence the initial part of a magnetisation curve is generally reversible in nature as shown in the diagram. It is only when the domain wall which contains those spins aligned in the field direction, crosses over a pin that removal of the field will not result in the domain wall returning to its initial position. At that point the change in magnetisation M_{REV} becomes irreversible and the sample, after exposure to such a field, will exhibit a relatively small but finite remanence. This process is shown schematically in Fig. 2.8 which is a reproduction of original experimental data.

As a larger field is applied the domain wall continues to sweep across the sample crossing many pins. As it does so the central portion of the magnetising curve exhibits irreversible behaviour and removal of fields in that range will show a growth in the resulting remanence M_{IRREV}. As a large field is applied such that the material approaches saturation the domain wall becomes squeezed against the edges of the sample itself and a large field is then required to induce rotation of the spins in the last domains. Hence domain wall motion is a very easy or soft process compared with rotation of the final small domains out of the sample.

The presence of the demagnetising field will result in the final few percent of the approach to saturation in the magnetisation again becoming reversible. On removal of the field the domain walls will start to relax back towards the initial starting configuration. However the reversal once the field is removed, is now driven by the magnetostatic energy due to the sample shape demagnetising field. As the magnetisation reduces so does the sample shape demagnetising field meaning that the domain walls may not have sufficient magnetostatic energy to cross over some of the stronger domain wall pins in the material. This then results in a significant remanence (M_r) requiring a reverse field of the order of the coercivity to return the magnetisation to zero. Hence it is always the case even in near perfect single crystals and as we will see in the case of single domain particles such as those found commonly in granular magnetic films, that the magnetisation is composed of two parts given by eqn 2.22.

$$M_{TOT} = M_{REV} + M_{IRREV} \qquad (2.22)$$

Of course in eqn 2.22 the term M_{REV} itself has two components both at very low fields and on the final approach to saturation. It is also interesting to note that a measure of the strength of the domain wall pins can be obtained by undertaking a measurement where a field is applied and subsequently reduced to zero and the remanence measured and then progressively larger fields are applied and again the resulting remanence measured. Because of the very nature of domain wall pins, the pinning strengths that result are never single valued. The distribution of pinning strengths within a given sample can be obtained from what is known as an isothermal remanent magnetisation (IRM) curve whereas such information cannot be obtained from a hysteresis loop because at every point around the loop except at saturation, there is a combination of reversible and irreversible components. This concept of measuring such energy barriers and their distributions irrespective of their origins, is discussed in Section 3.6.

2.4   Single Domain Particles

Given that this text is primarily concerned with particulate and granular magnetic materials where the particle or grain size is generally less than 30 nm, it is important to consider the phenomenon of single domain particles because as the dimensions of the particle are reduced there is not necessarily the physical space to accommodate a domain wall. The existence of particles containing a single magnetic domain was postulated way back in the 1930s by Frenkel and Doefman. In the 1940s and 50s a number of attempts were made to calculate from first principles where a transition from multi domain behaviour to single domain behaviour would occur. These were variously estimated for the common ferromagnetic elements at about 15 nm by Kittel in 1946 but was subsequently revised to 60 nm by Kittel and Galt (1956) some ten years later.

Attempts were made to infer the existence of single domain particles by measuring the field required to saturate two systems of nickel particles one of which had an approximate average diameter of 29 nm and the other having a diameter of 8 μm where the former was thought to be single domain and the latter was believed to be multidomain. It was found that the smaller particles saturated in $\sim$500 Oe and that the larger particles required a field of $\sim$2 kOe consistent with calculations of saturation by domain wall motion (Kittel *et al.*, 1950).

Final confirmation of the existence of single domain particles was provided by Morrish and Yu (1956) who undertook a highly elegant experiment using a high sensitivity torque magnetometer and observed the predicted square loop behaviour when the spins in a very small particle reverse over an anisotropy barrier remaining parallel as they do so producing a perfectly square hysteresis loop. The result of Morrish and Yu is reproduced in Fig. 2.9 where it can be seen that there is a step in the hysteresis loop which they attributed to the fact that they had accidentally measured two single domain particles that had become stuck together presumably due to dipole-dipole interactions.

Of course the criteria for determining whether a particle contains a single magnetic domain or a domain structure complete with domain walls has nothing to do with the physical size of the particle. The formation of a single domain within a particle is

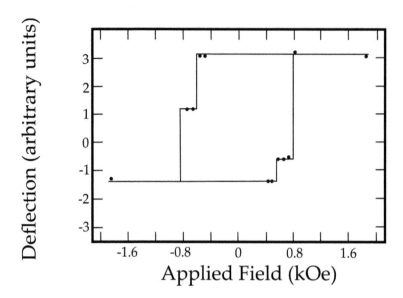

Fig. 2.9: Deflections in a torque magnetometer of an acicular micron-sized γ-Fe$_2$O$_3$ particle produced by applying an inhomogeneous magnetic field as a function of a previously applied magnetic field pulse. Reprinted figure with permission from [Morrish and Yu, Phys. Rev., 102, 670 1956] Copyright (1956) by the American Physical Society. Electronic format permissions: http://dx.doi.org/10.1103/PhysRev.102.670

determined by the difference in energy between the particle having a domain structure complete with a domain wall or whether it is energetically more favourable to form a single magnetic domain such that the spins remain parallel and are permanently saturated usually in an easy direction due to magnetocrystalline or shape effects, in the absence of an applied field. As we have seen the magnetostatic energy resulting from a particle being in a single domain state is given by

$$E_{ms} = \frac{1}{2} H_d M_s \tag{2.23}$$

The calculation of the critical size where the transition is likely to occur based on these energy considerations is extremely complex and is given in Chikazumi (1997). The result of this calculation gives that the critical size for single domain behaviour (d_{SD}) as

$$d_{SD} \approx \frac{(AK)^{1/2}}{M_s^2} \tag{2.24}$$

where A is the exchange stiffness constant. The exchange stiffness constant is a measure of the total exchange energy in the unit cell of a crystal lattice as opposed to that for two isolated spins given by eqn 2.11. Due to the effect of the lattice parameter it has units of ergs/cm. This approximate equation applies only when K is large and gives estimates lower than observed values. For materials with a moderate magnetocrystalline anisotropy e.g. Fe or Ni, this gives a critical size about one-tenth of the domain wall thickness. For high anisotropy for example in hcp Co, the domain wall energy is large and we get values of d_{SD} of about 5 to 50 nm. A more detailed calculation for the other two ferromagnetic metals gives similar values.

Basically the physics of eqn 2.24 indicates that if the exchange length is high so is the critical size because it takes a lot of energy to form a wall. If M_s is low then the critical size is large due to the M_s^2 term because the magnetostatic energy is low.

However in practice these calculations are found to be quite inaccurate. For example in many granular systems such as those found in magnetic recording tapes or even in hard disk media the particles or grains are deliberately grown so as not to be spherical. Under these circumstances it is generally found that the transition to the single domain state depends on the size of the smallest dimension of the particle. It is also worth noting that in many of the technological applications towards which this text is directed, the materials are either produced by chemical deposition or thin film techniques such as evaporation and sputtering, so grain sizes are generally below and often far below, 15 nm and hence at that size a single domain state is inevitable.

The only way that such a single domain particle can reverse its direction of magnetisation is by the spins rotating over one (or more) anisotropy energy barriers arising due to either magnetocrystalline or shape effects which can be a relatively magnetically hard process depending on the value of the anisotropy barrier which is to be overcome.

The first paper to address this topic is the pioneering work of Stoner and Wohlfarth (1948) who undertook a highly detailed quantitative study of the nature of reversal despite the fact that the many calculations were undertaken by Peter Wohlfarth using a mechanical hand driven adding machine! For simplicity they restricted their calculations to a uniaxial particle such as would occur in hcp cobalt or would occur in a particle of Fe or Ni having a shape anisotropy resulting from a particle elongation of >10%. A similar value of elongation to create uniaxial behaviour also occurs in magnetite and many of the other ferrites. In fact given that we are talking about particles that are typically of the order of 10 nm it is almost impossible to grow such particles which do not have an elongation of this order. Hence the theory is broadly applicable.

For clarity it should be emphasised that whilst the work of Stoner and Wohlfarth is perhaps the most highly cited paper in the field of particulate and granular magnetism, the calculations have certain limitations of which the reader should be aware. These are as follows:

1. The calculation does not include any element of thermal activation and hence applies strictly only at zero degrees Kelvin.

2. The assumption is made that the spins within the single domain particle are not only parallel but remain parallel, throughout the whole reversal process. This is known as coherent reversal. As we will see this is not always the case.

3. The calculation applies only to a system with uniaxial anisotropy but as will be described, a theory based on the principles of Stoner and Wohlfarth has been undertaken for systems with cubic anisotropy.

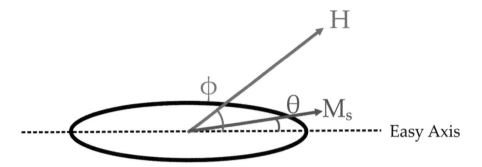

Fig. 2.10: Schematic of the classic Stoner-Wohlfarth particle.

Consider now the elongated particle shown in Fig. 2.10. Obviously such a single domain particle would have an easy axis along its long axis. In zero field the magnetic moment would be pointing along the easy axis. Due to the application of the applied field H at some other angle ϕ, the moment has moved away from the easy axis to an angle θ.

Firstly we can write down the energy of a single domain particle in such a condition which has two terms the first of which is the anisotropy energy resulting from the fact that the magnetic moment is not lying in an easy direction. The second term commonly called the Zeeman energy, arises because the magnetic moment is not aligned with the field.

$$E = KV sin^2\theta - M_s V H cos(\theta - \phi) \tag{2.25}$$

where the angles are defined in Fig. 2.10. Note that the $sin^2\theta$ variation in the first term arises because the energy must be the same and positive for a rotation in either direction. M_s will align in a direction such that the energy is minimised. Hence

$$\frac{\partial E}{\partial \theta} = KV sin2\theta - M_s V H sin(\theta - \phi) = 0 \tag{2.26}$$

For simplicity we now consider the case where the easy axis of the particle is aligned with the field, i.e. $\phi = 0$. This gives the minimum energy at $\theta = 0°$ and $\theta = 180°$ and a maximum at $\theta = 90°$ as expected.

Hence as the field is applied along the easy axis the moment remains in the easy direction until the Zeeman energy overcomes the anisotropy energy when a sudden reversal occurs giving rise to the square loop observed by Morrish and Yu (1956) shown in Fig. 2.9. In this unique case the reversal occurs at a field H_K

$$H_K = \frac{2K}{M_s} \tag{2.27}$$

known as the anisotropy field because it is the field that overcomes the anisotropy or the field that rotates the moment to 90° away from the easy axis. Hence the particle behaves as a two-state system separated by an energy barrier (ΔE) when a field is applied

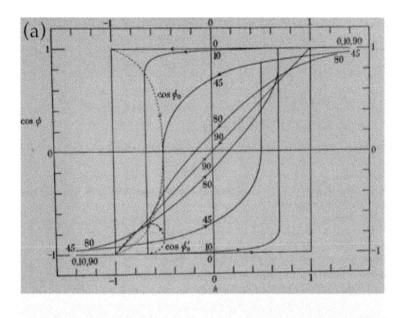

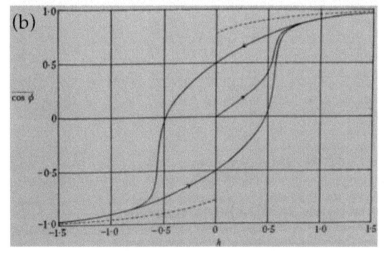

Fig. 2.11: (a) Stoner-Wohlfarth hysteresis loops as a function of alignment for individual particles and (b) Stoner-Wohlfarth hysteresis loop for a system of particles with easy axis randomly oriented (Stoner and Wohlfarth, 1948).

$$\Delta E = KV \left(1 - \frac{H}{H_K}\right)^2 \tag{2.28}$$

hence an applied field serves to reduce the barrier in proportion to H_K. In zero field, eqn 2.28 reduces to $\Delta E = KV$. The predicted square loop is labelled $\theta = 0$ in Fig. 2.11(a) taken from the original publication by Stoner and Wohlfarth (1948).

The mathematical analysis for the non-aligned case is quite complex and can be found in the original paper by Stoner and Wohlfarth (1948). However the physics of what occurs is quite clear. Referring to Fig. 2.10, when the easy axis along which the magnetic moment m $(m = M_sV)$ lies, once a field is applied there exists a torque τ on the moment given by

$$\tau = V\mathbf{M_s} \times \mathbf{H} = \mathbf{m} \times \mathbf{H} \tag{2.29}$$

This gives rise to a moment rotation away from the easy axis. As can be inferred from Fig. 2.10 this is a reversible change in the magnetisation along the easy axis direction since if the field is removed, the moment returns to the easy axis. However for each angle there exists a critical field at which the moment switches irreversibly towards the opposite easy axis direction.

For the case of an easy axis misaligned with the field by as little as 10°, at a critical field equal to $0.7H_K$ a sudden dramatic switch occurs and then a gradual alignment of the magnetic moment into the reverse field direction as shown in Fig. 2.11(a). It is noteworthy that the reduction in the switching field for an offset of only 10° reduces the switching field by 30%. This dramatic reduction shows the effect of the torque on the moment for any non-aligned angle. We also show the case when the magnetic field is applied at 90° to the easy direction and under these conditions the magnetisation becomes completely reversible with no switching observed.

In Fig. 2.11(b) we show the predicted hysteresis loop for a system where the easy axes are distributed at random in 3-D in relation to the direction of the applied field. Under these circumstances the resulting remanence is $0.5M_s$ as opposed to the aligned case where the remanence is equal to M_s. The observed coercivity is now equal to $0.479H_K$ where 0.479 is the 3-D average of $\cos\theta$. For the randomly oriented easy axes a large fraction ~50% of the change in magnetisation with field is reversible in nature. For the randomly oriented case the value of the anisotropy field is no longer $2K/M_s$ but is equal to $0.96K/M_s$ (Luborsky, 1961).

Whilst the Stoner-Wohlfarth theory was originally derived for particles exhibiting uniaxial anisotropy, using exactly the same principles the case for a material exhibiting cubic anisotropy was developed by Joffe and Heuberger (1974). However their work was also applicable only at $T = 0$ K and for a single particle size. This work was subsequently extended by Walker *et al.* (1993) using a computer simulation method and in both cases it was found that for cubic anisotropy the energy barrier was reduced according to

$$\Delta E = \frac{KV}{4} \, for \, K_1 > 0 \;\; e.g. \; Fe \tag{2.30}$$

$$\Delta E = \frac{KV}{12} \, for \, K_1 < 0 \;\; e.g. \; Ni \tag{2.31}$$

However because of the availability of six easy axes particularly for the case of $K > 0$, this results in a significantly increased remanence to saturation ratio (M_r/M_s) for the randomly oriented case at 0 K of 0.83. Of course whilst the cubic anisotropy case is important for completeness, generally the low value of cubic anisotropy constants means that at normal temperatures nanoparticles of cubic materials are generally superparamagnetic until they approach the size at which the transition to multi domain behaviour occurs. The one possible exception to this rule is cobalt ferrite which despite having a cubic spinel structure, has a significant magnetocrystalline anisotropy of 2.3×10^6 ergs/cc due to the anisotropic distribution of the Co and Fe ions within the lattice. At very low temperatures particles of this material have been found to exhibit the squareness value given above (Charles *et al.*, 1988).

2.5 Incoherent Reversal

As indicated above, the Stoner-Wohlfarth model and others deriving from it, make the assumption that the spins in small particles remain aligned parallel within the nanoparticle creating a single magnetic moment that then rotates against an anisotropy energy barrier. During the 1960s there was a significant effort led by workers at the then General Electric Research Center in Schenectady, New York, to use the potential of shape anisotropy in small particles of mainly Fe and Co to create materials that would have a very high coercivity suitable for use as permanent magnets. This group produced many papers but notably, work by Luborsky (1961) established that the coercivity predicted by the shape of the particles could not be realised. Importantly the particles were grown by electrodeposition into a liquid mercury cathode in the presence of a magnetic field which induced the growth of needle shaped particles. Measurements of the coercivity from very small sizes of a few nanometres, indicated that, as expected, the coercivity increased, but as shown in Fig. 2.12, at sizes >20 nm it was observed that the coercivity began to fall having a maximum coercivity of about 1 kOe where theory would have predicted 10 times this value. At a size of ~20 nm it would be expected from theory that the particles would still be in a single domain state and hence there was a contradiction between experimental data and theoretical predictions.

From electron microscopy images it was noted that the particles seem to have a shape like that of a peanut due to the coalescence of several particles in the liquid metal to produce the elongated structures. This apparent contradiction in the experimental data and the particular appearance of the particles, led to the realisation that perhaps the particles did not rotate coherently, i.e. with all spins oriented parallel. Jacobs and Bean (1955) described the particles reversing as a chain of spheres where the spins in the particles forming this apparent chain of spheres rotated in opposite directions as that would minimise the external field and hence the magnetostatic energy, during the reversal process. This reversal mode was often called the Jacobs and Bean mode, the chain of spheres mode or known as fanning, as the spins undergoing rotation in this manner appeared like the blades of a fan.

Figure 2.13 shows a schematic of such a chain of spheres and careful analytical calculations showed that this reversal mechanism best predicted the observed limit to the coercivity. Other variations again where parts of the particle underwent reversal via an incoherent mechanism where the spins did not remain parallel, were also developed around the same time which included phenomena such as curling and buckling of the moments details of which can be found for example in Cullity and Graham (2008).

As described in Chapters 3 and 6, in later years it became known that all these analytical descriptions of incoherent reversal were of course approximations and that the phenomena of incoherent reversal could be induced by surface roughness on elon-

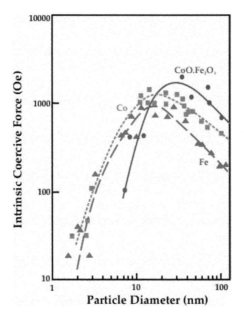

Fig. 2.12: Coercivity as a function of size for particles deriving their coercive force principally from shape anisotropy Reprinted from Luborsky J. Appl. Phys. 32 S171 (1961), with the permission of AIP Publishing.

gated particles and that the description for example of a chain of spheres was far too simplistic to describe the actual spin configurations during incoherent reversal which required the development of computer models in the 1980s to show the exact mechanisms (Schabes and Bertram, 1988). None-the-less all these early and indeed later works, showed that, particularly in elongated or non-perfectly spherical particles above about 20 nm in diameter, the assumption of coherent reversal giving rise to the energy barriers provided by the Stoner-Wohlfarth model did not apply.

Fig. 2.13: Schematic of fanning reversal.

2.6 Exchange Coupling in Granular Films

One further complexity in any granular material is the existence of coupling effects between the grains in the material. In all particulate and granular materials there will always be free poles on the surface of the particles giving rise to dipole-dipole interactions. This topic is discussed in Section 3.8. However because such interactions do not give rise to the formation of domain structures as in a regular solid, here we consider the effect of exchange coupling in granular films.

Because the range of direct exchange coupling in solid materials is very short extending out only over one or two atoms, the fact that particles are quite close together does not allow for the development of significant direct exchange coupling. However there is a special case of metallic granular films where the proximity of the grains to each other in a material, produced by electrodeposition, sputtering or other similar deposition techniques, allows for the movement of conduction electrons through the material and despite the discrete nature of the grains, such films can be quite good conductors of electricity. This allows for the existence of a Ruderman, Kittel, Kasuya and Yosida (RKKY) interaction which can be very strong (Ruderman and Kittel, 1954; Kasuya, 1956; Yosida, 1957).

The principle of the RKKY interaction is that conduction electrons passing through a grain oriented in one particular direction will result in the polarisation of the electrons with many passing through the same particle. The polarisation of this electron stream when entering a neighbouring particle or grain then results in the alignment of the moment of that (usually) single domain particle. In this way from the consideration of the magnetic behaviour of such a thin film, the material itself becomes quasi-continuous allowing for very strong intergranular interactions to be developed. Such interactions and the control of them has become a major issue in the development of thin film materials, but in particular the recording layers in hard disk drives. This has only been overcome in the last few years by the development of co-sputtering of an insulator which segregates the grain boundaries thereby preventing the propagation of an RKKY type interaction. This topic is addressed in some detail in Section 7.4.

Due to the presence of RKKY interactions for example in a thin film of say pure iron, pure cobalt or an Fe-Co alloy, means that even though the film is composed of small entities, that domain structures can form in metallic thin films. In this case the reversal mechanism in such films behaves in a directly analogous manner to that of a continuous thin film or a single crystal. The only significant difference that occurs is that any reverse domain must first nucleate and because the grains are discrete, this involves a nucleation via a Stoner-Wohlfarth type mechanism over an energy barrier. That is also true for a continuous material. However the presence of an infinite number of grain boundaries, defects and voids in such a film also leads to a very high density of domain wall pins at the boundaries of the grains. In particular any void between the grains can create a very strong domain wall pin. For this reason thin films of this type generally have a coercivity that is significantly larger than would be the case for a single crystal.

The presence of RKKY interactions in metallic granular thin films leads to another form of hysteresis. Due to the interactions and particularly where the material has low or moderate anisotropy, it is often the case that the resulting hysteresis loops are completely square i.e. $M_r/M_s = 1$ and a vertical variation through the switching region and coercivity occurs as shown in Fig. 2.14 for thin films of CoFe.

The basis of the squareness $M_r/M_s = 1$ is the RKKY interaction with the strong intergranular coupling maintaining the moments of grains which have sizes ~10 nm, aligned parallel. The reversal is initiated by a nucleation event perhaps due to a grain

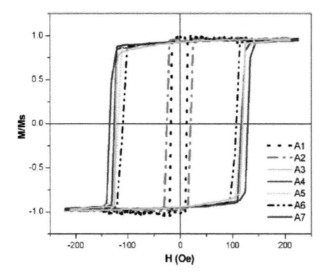

Fig. 2.14: Room temperature hysteresis loops for CoFe thin films deposited at different bias voltages (Vopsaroiu *et al.*, 2005).

that is weakly coupled to others or more likely near the edge of the sample where some roughness leads to a locally strong demagnetising field. The resulting reverse domain then propagates rapidly through the sample due to the RKKY interactions which overcome any anisotropies, impeded only by the sample shape demagnetising field which is zero in the plane of the film but may exist locally at grain boundaries.

Hence the measured coercivity is not a reflection of the intrinsic properties of the grains themselves. The reversal is dominated by the weak link reversal at the nucleation field H_n which is often equal to H_c.

The effects of edge roughness on H_n were studied in our lab some years ago and we found that any form of sample cutting induced low values of H_n. Similarly deposition through a mask had a similar effect due to the film being thinner near the edges. Using suitably sized substrates which were coated completely gave reproducible data (Hussain *et al.*, 2011).

2.7   Ferrimagnetic Nanoparticles

It is possible to produce nanoparticles of the ferromagnetic transition metals by a number of techniques but even when the particle size is relatively large i.e. above $1\mu m$, the particles are not oxidatively stable. In addition, finely divided metal particles of any kind are generally catalytic and therefore can be toxic. This is particularly true of nickel which is a known carcinogen. Hence for all applications in the field of magnetic nanoparticles it is necessary to use the ferrimagnetic iron oxides or ferrites as they are commonly known.

Based on the discussion of the origin of ferromagnetism in Section 5.2 obviously the mechanism of the ferrimagnetism in iron oxides must be different to that in metals. In the iron oxides there is a unique form of electrostatic coupling that gives rise to magnetic order. This phenomenon is known as the Superexchange Interaction. Consider now Fig. 2.15 which shows the electron orbitals from any transition metal coupling to oxygen. As shown in the diagram it is the 2p orbitals of the oxygen which are responsible for the chemical bonding. The bonding then takes place with a 3d orbital in the metal. As is clear from the figure, the p orbitals are quasi elliptical in shape and it is required that for the two electrons involved in the bonding that one should be spin up and other should be spin down. This imposes a similar restriction due to the Pauli Exclusion Principle, on the spin orientation of the 3d electrons which also now have to bond with a spin in the opposite direction to that in which the electron in the oxygen was orientated. As can be seen in the figure this imposes an antiparallel or antiferromagnetic order on the metal. Hence many oxides of the transition metals including MnO, CoO, NiO are antiferromagnets and as discussed in Chapter 4. NiO was the first antiferromagnet used in the production of spin valve sensors. Interestingly NiO is still the only antiferromagnetic oxide that has been used in an industrial application. The position regarding oxides of iron is far more complex but none-the-less the magnetic order in all cases derives from superexchange coupling.

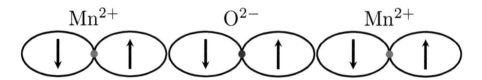

Fig. 2.15: Coupling of the 3d orbitals in transition metals to the 2p orbitals in oxygen.

Iron as an element has two valence states being Fe^{2+} and Fe^{3+}. However as we shall see there is also a strong influence of the crystallographic order of the iron oxides on the resulting ferrimagnetism. For example the simplest of the iron oxides is Fe_2O_3 known as hematite. When grown in one crystallographic phrase known as the α-phase this material is also an antiferromagnet. However when it is grown in a spinel crystal structure which is known as the γ-phase known as γ-Fe_2O_3, it is ferrimagnetic. This is purely because the Fe^{3+} ions are distributed differently within the crystal lattice so that perfect cancellation of the magnetic moments does not occur.

The full crystal structures of maghemite and particularly in the case of magnetite (Fe_3O_4), are extremely complex and whilst there are 3d images of the full crystal structure to be found in other textbooks and particularly on websites, these are very difficult to envision and even more difficult to interpret. For example in Fe_3O_4 there are 56 ions per unit cell. The larger oxygen ions which have a radius of about 0.13 nm are packed quite closely together in an fcc unit cell and the much smaller metal ions which have radii of about 0.075 nm occupy the spaces between them. The arrangement of the metal ions in relation to the oxygen come in two forms, the first being known as a tetrahedral or A site and the second is an octahedral or B site. These arrangements are shown in Fig. 2.16. This results in a spinel structure where the distribution of the metal ions between the two types of sites is unequal. For example in a normal spinel structure

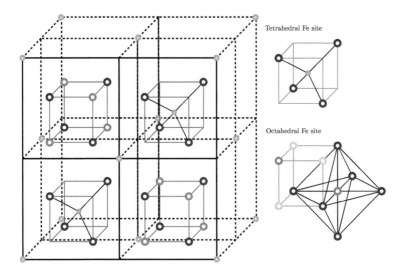

Tetrahedral Fe site

Octahedral Fe site

Fig. 2.16: Schematic of the crystal structure in magnetite.

there are 8 divalent ions on tetrahedral sites and 16 trivalent Fe ions on octahedral sites. For an inverse spinel structure there are 8 trivalent ions on the tetrahedral sites and a mixture of 8 trivalent Fe ions and 8 divalent ions on the octahedral sites. In fine particle form, particularly below say 20 nm, these ideal distributions which would occur in a single crystal, are rarely found leading to some crystallographic disorder. Whilst the material may still be ferrimagnetic in fine particle form it will not necessarily have the same magnetisation as would be the case in a bulk material.

The most commonly used ferrite in fine particle form and indeed for almost all the applications discussed in this text, the ferrite of choice is magnetite which nominally has the chemical formula Fe_3O_4 but in reality should be correctly written as $FeO.Fe_2O_3$. This ferrite has the highest magnetisation of any compound of this type at 480 emu/cc (92 emu/g at room temperature). Of course this value is much lower than that for Fe and Co which have saturation magnetisation values of 1700 emu/cc and 1440 emu/cc, respectively. Ni has a somewhat lower magnetisation at 480 emu/cc.

There are a plethora of other cubic ferrites which are to some degree magnetic but these are rarely used in nanoparticle form but some do find application as micron size particles in high frequency inductors because they are both magnetic and insulating and hence supress eddy currents at high frequencies. There are many other ferrites composed of a transition metal oxide with Fe_2O_3 that are known and form quite readily and perhaps the only other one that sees any significant usage is cobalt ferrite ($CoO.Fe_2O_3$). It is also the case that there are a few ferrites that form in a hexagonal structure such as barium ferrite ($BaO.6Fe_2O_3$) and also a similar structure formed by strontium ferrite. These have room temperature magnetisation values of around 72 emu/g and in particular find application because of the very high anisotropy resulting from the hexagonal structure. In particular in recent years barium ferrite has found application due to its anisotropy as a suitable particle having perpendicular anisotropy that can be used in very high density data tapes discussed in Chapter 7. However for all other applications of magnetic particles discussed here magnetite is the material of choice.

As mentioned above it is the non-ideal crystal structure of magnetite that occurs in nanoparticles that reduces the saturation magnetisation. Furthermore unless nanoparticles of magnetite are kept in an inert oxygen free atmosphere the particles generally oxidise through slowly from Fe_3O_4 to γ-Fe_2O_3. This can be seen with the naked eye because magnetite is black in colour whereas γ-Fe_2O_3 is orange and for example many ferrofluids that contain these nanoparticles are observed to change colour over a relatively short period of time. This causes a reduction in the room temperature saturation magnetisation from the ideal 92 emu/g to about 80 emu/g. However because of the crystallographic disorder that will occur in either form, the saturation magnetisation of magnetite particles is generally around 70-75 emu/g.

There is one other major application of magnetite which is little known in the magnetism community. In a photocopier or a laser printer it is required to have a mechanism to lift the pigment out of its reservoir with some sort of magnetic brush so that it can be attracted electrostatically to the photo-sensitive cylinder on which the image to be copied has been imprinted. This is done using micron size magnetite particles which are now in a multi domain state. The toner itself consists of polymer spheres typically of about 10-20 μm which contain a mixture of carbon black and the magnetite particles which are also black. These are transferred to the paper and then heated to compose the image. However because of the amount of printing that is done in the world, this application consumes many tons of magnetite worldwide each year, most of which (hopefully) ends up in recycling.

3
Thermal Activation Effects

All physical systems unless they are at a temperature of zero Kelvin or very close to, are subject to thermal activation effects which generally lead to disorder or scattering of entities. The best example is perhaps the well known electron-phonon scattering that gives rise to the temperature variation of resistance in metals and semiconductors. In certain materials and notably magnetic fine particles, the effects of thermal activation can be particularly marked because thermal activation in a single domain particle can lead to the reversal in the orientation of the magnetic moment of the particle which is the sum of the atomic moments on the atoms within it. This leads to a number of macroscopic phenomena that affect basic measurements and the properties of all systems.

The effects are also critical for data storage where both the ability to switch a moment as quickly as possible and in particular, the long term storage of the information can be compromised by the effects of thermal energy. This obviously makes significant impacts on both the ability to write data to a magnetic system and for storage which may be required for a period stretching to many years. Hence an understanding of the physical origins of thermal activation and its consequences for single domain particles is critical.

Whilst the discussion in this chapter is presented in terms of fine particles alone, it is the case that the basic physics is applicable to all hysteretic magnetic materials. For materials whose reversal is based on domain wall motion the energy barrier which for single domain particles in zero field is given by $\Delta E = KV$, is simply replaced by the energy barrier derived from the strength of a domain wall pin. In a similar way that ΔE for a fine particle system is not single valued due to the particle size distribution, the same applies to domain wall pins whose strength is not uniform. Indeed the first detailed study of thermal activation of reversal was made on Alnico magnets by Street and Wooley (1949).

However the first direct observation of a frequency ($= 1/$time) effect on hysteresis as made by the Scottish scientist/engineer Sir (James) Alfred Ewing (1855-1935) at King's College Cambridge. Ewing developed the first molecular theory of magnetism and coined the term hysteresis. He also introduced the Japanese to the study and applications of magnetic materials.

3.1   Single Particle Activation

As we have seen in Section 2.1 a single domain magnetic particle exhibiting uniaxial anisotropy deriving from either magnetocrystalline or shape effects with the easy axes aligned in the direction of the applied field has an effective energy barrier to reversal given by

$$\Delta E = KV \left(1 - \frac{H}{H_K}\right)^2 \tag{3.1}$$

The energy barrier in zero field then reduces to

$$\Delta E = KV \tag{3.2}$$

It should be noted that whilst in principle, single domain particles used for information storage are in zero field, the presence of the bits of information which generally comprise several grains aligned in one direction, leads to a demagnetising field and hence the information is not in practice stored in zero field. This effect is particularly critical where information is stored perpendicular to the plane of a thin film or coating when the demagnetising field becomes equal to $H_d = -4\pi M_s$. This topic will be discussed in detail in Chapter 7.

The Nobel Laureate Louis Néel first addressed this problem as long ago as 1949 (Néel, 1949a) and indicated that the energy barrier for a very fine particle can be so small because of its volume dependence, that magnetisation reversal may occur by thermal activation alone. He described the effect as being a simple exponential law of the form

$$\frac{m(t)}{m(0)} = e^{(-t/\tau)} \tag{3.3}$$

where t is the measurement time and the relaxation time (τ) was given by an Arrhenius relation of the form

$$\tau^{-1} = f_0 e^{-\Delta E/kT} \tag{3.4}$$

i.e.

$$\tau^{-1} = f_0 e^{-KV(1-H/H_K)^2/kT} \tag{3.5}$$

where f_0 is an attempt frequency analogous to the Larmor precession frequency in the anisotropy field H_K. The value of f_0 was first calculated for iron by Kneller (1964) who obtained a value of $f_0 = 10^9 \ s^{-1}$. This value is commonly used for almost all materials but there have been measurements that show that for different materials the value often varies in the range from 10^9 to 10^{13} (Dickson *et al.* 1993, Xiao *et al.* 1986). Equation 3.4 indicates that τ is the time taken for the magnetisation of an assembly of non-interacting identical particles to decay from an initial value to $1/e$ or 37% of its initial value. Note here that these equations refer to the aligned case based on eqn 3.1 rather than a random orientation but the $1/e$ variation still applies.

The variation of the relaxation time with particle diameter for a system of cobalt nanoparticles having a value of anisotropy constant $K = 2 \times 10^6$ erg/cc at a temperature $T = 290 \ K$ and assuming $f_0 = 10^9 \ s^{-1}$ is shown in Fig. 3.1. The value of K used here is typical for that which has been measured for cobalt particles produced by the decomposition of dicobalt octacarbonyl in the presence of a surfactant which results in $\sim$10 nm particles in an fcc phase with shape anisotropy which can form ferrolfuids. It should be noted that in this work the ferrofluid was frozen in zero field and hence in that case the easy axes of the particles would be randomly orientated. This means

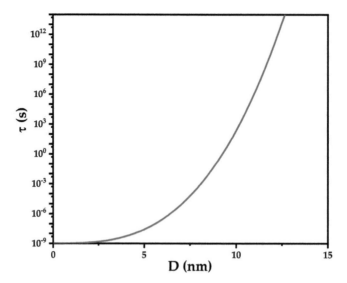

Fig. 3.1: Relaxation time in zero field as a function of particle size for Co nanoparticles with $K = 2 \times 10^6$ ergs/cc.

that the value of K measured and used for this illustration is an effective value of the anisotropy constant for a randomly oriented system and the shape anisotropy makes the value temperature independent.

As can be seen from Fig. 3.1 the variation of the relaxation time is extremely rapid due to the exponential nature of eqn 3.5 and the fact that the relaxation time depends on the volume and hence on D^3. This rapid variation generally leads to a sharp cut off at a given temperature in the stability of a magnetic moment in zero field as compared to the situation where the moment oscillates rapidly switching between two easy axes or being completely disorientated. This phenomena leads to the concept of superparamagnetism and the phenomena of a magnetic moment in a single domain particle being blocked or stable as discussed in Section 3.2.

From Fig. 3.1 it can be seen that for very small particles $D < 9$ nm the relaxation time after the removal of an applied field is of the order of milliseconds and time dependent phenomenon will not be observed using conventional magnetic measurement techniques such as a vibrating sample magnetometer with a measurement time of 100 s. In contrast for slightly larger particles ($D > 11$ nm) the relaxation time is of the order of 10^6 seconds or 11 days, and the magnetisation will then appear stable. Hence it is possible to define a critical diameter at which the magnetisation becomes stable. If the value of the measurement time is set for example to 10^2 seconds, which is of the correct order for a measurement done using a vibrating sample magnetometer, then the value of this critical diameter (D_p) equals 10 nm at 293 K and is insensitive to small changes in the relaxation time. For particles of diameter $D < D_p$ the magneti-sation will spontaneously reverse from one easy direction to another in the absence of an applied field. If a field is applied it will tend to align the moments whereas thermal energy will tend to disalign them. This is the behaviour of a normal paramagnet with the exception that the magnetic moment per particle, $m = M_s V$, is many thousand Bohr magnetons compared with only a few for an ordinary paramagnet. Accordingly, Bean and Livingston (1959) termed the magnetic behaviour of these particles super-paramagnetism.

3.2 Superparamagnetism and Blocking

For most real fine particle systems the orientation of the easy axis of magnetisation is random and hence in this case the Langevin function $L(\alpha)$ applies. It should be noted that for a very heavily aligned system which is in essence a two-state system, the Brillouin function may be more appropriate.

Superparamagnetic behaviour is characterised by two specific magnetisation effects based on the Langevin function given by

$$L(\alpha) = coth(\alpha) - \frac{1}{\alpha}; \ \alpha = \frac{mH}{kT} \tag{3.6}$$

These are as follows:

1. The magnetisation curves exhibit no hysteresis i.e. the coercivity and remanence are both zero.
2. Based on the form of the Langevin function, magnetisation curves measured at different temperatures superimpose when the magnetisation is plotted against field over temperature (H/T). The absence of hysteresis is often taken to be indicative of superparamagnetic behaviour but in fact it is equally important to ensure the superposition of magnetisation curves plotted against H/T to have true superparamagnetic behaviour.

A criterion to define the critical diameter (D_p) in zero field can be obtained by substituting a value for the measurement time into eqn 3.5 and using the value of f_0 obtained by Kneller (1964). Using a measurement time of $t = 100$ s leads to the criterion (Bean and Livingston, 1959)

$$KV_p = 25kT \tag{3.7}$$

which in turn leads to an equation for the critical diameter in zero field

$$D_p = \left(\frac{150kT}{\pi K}\right)^{1/3} \tag{3.8}$$

Equation 3.8 shows that D_p is temperature dependent and hence for particles of a constant size there exists a temperature T_B below which the magnetisation becomes stable. Such particles are generally described as being blocked and T_B is the blocking temperature. Figure 3.2 shows the variation of the critical diameter for superparam-agnetic behaviour with temperature using the same parameters as used for Fig. 3.1.

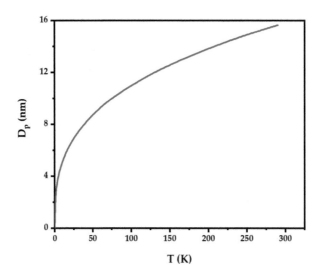

Fig. 3.2: Critical size for superparamagnetic behaviour in zero field as a function of temperature for Co nanoparticles with $K = 2 \times 10^6$ ergs/cc.

Again a rapid variation particularly at low temperatures is observed. However it should be noted that the value of D_p for any system is critically dependent on the measurement time. Modern instrumentation, particularly in use with vibrating sample magnetometers, can allow for very rapid sweep rates of the magnetic field so the criterion of setting D_p according to a measurement time of 100 s may not apply. Critically if a measurement is made using an AC magnetic field then the value of D_p can vary significantly. This effect is of particular significance for example in information storage both for writing of data which is undertaken at very high frequencies and for the long-term stability of stored information where the effective measurement time must be a number of years. It is also of critical importance to understand the time and hence the frequency dependence of the critical diameter when considering other measurements and phenomena such as that of magnetic hyperthermia which is discussed in Section 6.5.

A further consequence of the phenomenon of blocking in fine particles is that such particles show hysteresis possessing both a coercivity and remanence and behave as typical ferromagnets. Those particles of diameter significantly less then D_p will show superparamagnetic behaviour following a Langevin function. Most significantly particles having a diameter $D \sim D_p$ will exhibit significant time dependent behaviour which is often not taken into account. For such systems it is important that the sweep rate of a DC field measurement or the frequency of an AC measurement is taken into account carefully and kept constant when a range of samples are to be measured. Hence time and frequency effects on the measurement particularly of the coercivity, in such systems have been observed.

The discussion above is the case where a particle is in zero applied field to simplify the analysis and hence we must now consider the effect of an applied field on thermal activation. Again for simplicity we will consider the case where the easy axes of the uniaxial particles are aligned and the moments are saturated by an applied field H. If a reverse field $-H$ is now applied to the system then the energy of each particle is given by

$$E = KV \sin^2\theta + HMV \cos\theta \qquad (3.9)$$

where θ is the angle between the moment and the $+H$ direction. Equation 3.9 is a special case of eqn 2.25 with $\phi = 0$. The energy barrier to magnetisation reversal is now given by the difference between the maximum and minimum values of eqn 3.9. Thus a similar analysis to that in Chapter 2 gives

$$\Delta E = KV \left(1 - \frac{HM_s}{2K}\right)^2 \qquad (3.10)$$

This equation should be compared with $\Delta E = KV$ for $H{=}0$ and hence the application of a field reduces the barrier to magnetisation reversal. When a field is applied which reduces the value of ΔE to $25kT$ the magnetisation will reverse in 100 seconds as we are using a consistent measurement time. This field is then the coercivity of the particle given by

$$KV \left(1 - \frac{H_c M_s}{2K}\right)^2 = 25kT \qquad (3.11)$$

and hence the coercivity of a single particle is then given by

$$H_c = \frac{2K}{M_s}\left(1 - \left[\frac{25kT}{KV}\right]^{1/2}\right) \qquad (3.12)$$

Hence given that the term $2K/M_s$ is the anisotropy field, it is clear that the effect of thermal energy reduces the switching field i.e. the coercivity, is dependent upon the square root of the temperature and the volume of the particle. At very low temperatures of course H_c tends to the value of H_K which defines the anisotropy field as the value of the coercivity at $T{=}0$. We can now define the coercivity relative to the value of the anisotropy field in terms of the critical diameter for superparamagnetic behaviour

$$\frac{H_c}{H_K} = \left(1 - \frac{D_p}{D}\right)^{3/2} \qquad (3.13)$$

3.3   Particle and Grain Size Analysis

The preceding discussion is applicable to non-interacting, monodispersed, uniaxial particles whereas in practice all systems consisting of fine particles or granular thin films will contain a distribution of particle volumes, $f(V)$. In addition there will also be a distribution of particle shapes leading to a distribution of shape anisotropy constants, $h(K)$. The distribution of particle shapes will be most significant for materials whose magnetocrystalline anisotropy is relatively low e.g. magnetite (Fe_3O_4) and other iron oxides and metallic systems where the anisotropy is low in cubic materials. This will include e.g. iron, cobalt and nickel but in the case of cobalt only the case where the crystal structure is fcc rather than hcp which is commonly found to be the case in fine particles. To further complicate things, this will not apply to cobalt thin films such as those used in the hard disk of recording media where the availability of epitaxial growth ensures that the grains in the film are in an hcp structure. In that case the effects of shape anisotropy are more limited. The other common exclusion to the rule is barium ferrite where in addition to having a relatively low magnetisation thereby limiting the effects of shape anisotropy, the anisotropy of this oxide and other hexagonal oxides e.g. those containing strontium, is very large. In these cases the effect of shape anisotropy is a perturbation on the magnetocrystalline term but the hexagonal structure ensures that they remain uniaxial particles.

However in practice, a common technique employed is to consider the distribution of diameters and to assume an effective value of the anisotropy constant and simply consider the distribution of particle volumes.

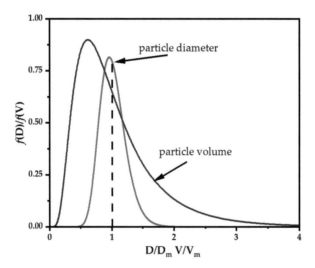

Fig. 3.3: Particle and volume size distributions for a sample with an average particle size of 10 nm and a standard deviation of 0.2.

In almost all particulate and granular materials it is found that the distribution of diameters generally follows the lognormal distribution function shown in Fig. 3.3. This asymmetric distribution is a direct result of the mechanisms of particle growth which generally occurs via a nucleation process either in some kind of chemical solution or when thin films are produced by sputtering, with the coalescence of several atoms to form a nucleus. As more material is deposited this leads to a growth phase where each nucleus will grow uniformly. However in practice this does not occur due to statistical effects and the result is that there will be initially at this point, a distribution of diameters that should be Gaussian in nature.

However as the growth process proceeds the well-known phenomena of Ostwald ripening occurs whereby smaller particles are absorbed by bigger particles which then fuse to make a single entity or crystallite due to the very high surface energy of the smaller particles. This process then favours the production of larger grains unless some inhibiting mechanism occurs such is the case where particles are grown in solution incorporating the use of surfactants which are chemicals that bind to the particle surface as the growth proceeds. The theory of how this results in a lognormal distribution was described by Granqvist and Buhrman (1976).

As we have seen the properties of particulate and granular magnetic materials are controlled by the particle volume. That means that even where a chemical process is used and a Gaussian distribution of diameters results; when the distribution of volumes is considered, which of course depends on D^3, then a distribution closely following the lognormal distribution exists due to the D^3 factor.

For purposes of simplicity therefore we will restrict this discussion to a lognormal distribution of particle diameters which also leads to a lognormal distribution of particle volumes. The lognormal distribution function is defined as being a distribution such that the parameter $ln(D)$ follows a normal or Gaussian distribution. Hence the formula for the distribution is of the form

$$f(ln(D))d(ln(D)) = \frac{1}{\sqrt{2\pi}\sigma}exp\left[-\frac{\left(ln(D) - \overline{ln(D)}\right)^2}{2\sigma^2}\right]d(ln(D)) \qquad (3.14)$$

Clearly this distribution is difficult to manipulate and use because it requires that the interval of measurement is evenly spaced in $ln(D)$. However a simple transformation of variables can be used to produce a function such that the interval is linear in D as shown in eqn 3.15. Whilst this function is simple to use there are a number of features that are worth discussing. These are the presence of the term $1/D$ in the normalisation factor which leads to the higher end of the distribution following a $1/D$ curve quite precisely. Furthermore whilst the interval has been transformed from $ln(D)$ to D, it remains the case that the resulting standard deviation is the standard deviation of $ln(D)$ and not that of D itself. The median diameter is given by $e^{\overline{ln(D)}}$ where $\overline{ln(D)}$ is the mean value of $ln(D)$ and does not coincide with the mode or peak in the distribution. The standard deviation of $ln(D)$ can be converted to a standard deviation of D but it does not have the same physical description in terms of the lognormal distribution function curves.

$$f(D)dD = \frac{1}{\sqrt{2\pi}\sigma D}exp\left[-\frac{\left(ln(D) - \overline{ln(D)}\right)^2}{2\sigma^2}\right]dD \qquad (3.15)$$

By way of illustration Fig. 3.3 shows a lognormal distribution function for an imagined distribution where the median diameter D_m is taken to be 10 nm and the standard deviation of $\ln(D)$ σ is taken to be 0.2. Also shown on the same graph is the resulting distribution of the particle volumes which are the most significant for magnetic systems. The median values are marked as D/D_m and V/V_m showing the difference between the two cases in terms of the peak. It is also worthy of note that the distribution of volumes is much wider because considering the variable D to be raised to the third power results in the standard deviation being three times larger for volume than is the case for diameter. For all calculations of magnetic properties of particulate and granular systems it is of course the distribution of volumes that must be considered and not that of diameters. The diameter of a particle having the median volume (D_{vm}) is related to the median diameter (D_m) by

$$D_{vm} = e^{3\sigma^2} D_m \tag{3.16}$$

and

$$\sigma_V = 3\sigma_D \tag{3.17}$$

For the case of thin films particularly where columnar growth has occurred, the grain volume is measured by measuring the area of each grain multiplied by the film thickness. This is generally valid because the thickness of films grown by sputtering is highly uniform and known to high accuracy via the use of thickness monitors based on piezoelectric resonance. In this case the distribution comes from the area of the grains (A) and

$$\sigma_A = 2\sigma_D \tag{3.18}$$

This also represents the standard deviation of $\ln A$.

With the availability of modern graphic packages it is relatively simple to fit eqn 3.15 to a set of experimental data. However as the fit involves the two parameters D_m and σ (of $\ln D$) the resulting solution may not be unique. Also any error in σ will be amplified where σ_V is required.

The best technique is to construct a simple spreadsheet from the data following eqn 3.14 to find $\overline{\ln D}$ which gives D_m via

$$D_m = exp\left(\overline{\ln D}\right) \tag{3.19}$$

As the distribution of $\ln D$ is Gaussian the value of the variance σ^2 is given by

$$\sigma^2 = Mean\ of\ the\ squares\ -\ Square\ of\ the\ mean \tag{3.20}$$

and hence

$$\sigma = \left[\frac{1}{N}\sum_{i=1}^{N}(\ln D_i)^2 - \left(\frac{1}{N}\sum_{i=1}^{N}\ln D_i\right)^2\right]^{1/2} \tag{3.21}$$

where N is the number of particles or grains measured. The curve to fit the experimental data can then be obtained directly from eqn 3.15.

In order to calculate the magnetic properties of a real system it is generally required to integrate a measured distribution function usually between limits created either by the available switching field or the effects of temperature. In modern times this is generally quite easy to do using commonly available graphics packages such as Origin Lab (https://www.originlab.com/).

It is also the case that many particulate and granular systems readily allow the use of transmission electron microscopy to obtain high resolution images of the particles or grains. This is particularly true of the discrete particles that can be quite easily dispersed in a suitable liquid using a surfactant which in the case of magnetic materials is commonly oleic acid which bonds readily to both oxides and metals but is not compatible with water so a solvent such as heptane has to be used to coat the grid which is then left to dry.

In the case of thin films it is generally the case that the substrate has to be thinned by a combination of grinding with an abrasive material in the form of sandpaper followed by some form of plasma etching once the overall thickness of the substrate has been reduced. Figure 3.4 shows transmission electron microscope (TEM) images of a ferrofluid containing magnetite particles and of a thin film of an exchange bias system thinned as described above. In each case the resulting measured particle size distribution is shown in terms of the grain diameters. Note that the values of σ are those of lnD.

For the case of the nanoparticles shown in Fig. 3.4(a) it is relatively easy to identify the boundaries of the particles. However for the thin film sample shown in Fig. 3.4(b) this is not the case. In the case of thin films the contrast generated by the grains comes in two forms: the first of these is mass-thickness contrast depending on the atomic number of the atoms in the grains as the scattering is elastic nuclear scattering (i.e. Rutherford scattering). The thickness of the film also contributes to the contrast as a thicker film will generate more scattering events.

The second source of contrast is from electron diffraction. For granular thin films generally not all grains have the same crystallographic texture. Some of the grains will be oriented so that they meet or almost meet the Bragg diffraction condition

$$n\lambda = 2dsin\theta \tag{3.22}$$

where λ is the wavelength of the electrons, d is the interplanar spacing of the crystallite in the grain and θ is the Bragg angle. These grains appear black or very dark in the image in Fig. 3.4(b). Other grains not meeting the Bragg condition appear in various shades of grey due to mass-thickness contrast.

For the grains that satisfy the Bragg condition it is easy to see the grain boundaries and measure their size. In our laboratory it is only these grains that are measured. Alternatively if dark field (diffraction plane) imaging is used, the grains that meet the Bragg condition appear white and are even easier to measure. The authors are grateful to Prof. J. N. Chapman of Glasgow University (private communication) for suggesting this technique which we believe is not published.

There are a number of techniques available for the measurement of grain diameters such as the commonly used software ImageJ. However there is always a requirement for the measurements to be done by hand due to variations in particle shape. Hence the

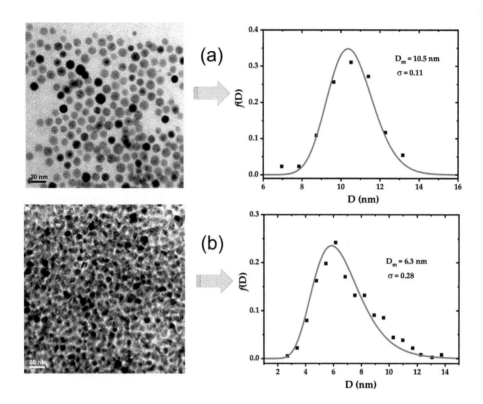

Fig. 3.4: TEM image for (a) a ferrofluid and (b) an exchange bias thin film sample and corresponding particle/grain size distributions.

methodology can be described as an equivalent circle method. Whilst this is possible using software based analysis, in our experience a much better method is to use a light box with a variable aperture calibrated to convert the equivalent circle diameter into a computer readable value. The system that we have used is based on a very old device called a Zeiss Particle Size Analyser with which it is possible to measure up to 500 particles in an hour which is much quicker than using software such as ImageJ.

The next question that arises is how many particles need to be measured. We have done an in-house study of this and found that the measurement of as few as a hundred particles is capable of giving the median diameter to an accuracy of below 10%. To achieve this it must be ensured that no bias occurs in the selection of particles to measure. The best technique is to draw a grid on the image with squares of say one or two centimetres and then to measure every particle visible in any square taken at random but then ensure that the next square within which particles are measured is adjacent to the first one. This removes any human bias that might occur.

However when calculating the magnetic properties of real particulate and granular systems and when the requirement is to calculate the resulting properties, the standard deviation becomes a far more critical parameter. From our in-house study we have found that it is necessary to measure at least 500 particles to obtain the standard deviation to a similar resolution to that obtained for the median diameter. This also has the advantage of increasing the resolution of the value of the median diameter.

3.4 Coercivity and Remanance

We can now address the effect of a particle size distribution on the key parameters giving rise to magnetic hysteresis. Figure 3.5 shows the same lognormal distribution function in terms of diameter as that shown in Fig. 3.3 with the exception that values of the critical diameter D_p at a range of temperatures has been shown using the same conditions as used in Fig. 3.2. Because of the variation of D_p with temperature the fraction of the distribution which contains blocked particles will decrease as the temperature increases. This gives rise to a very simple relationship between the remanence and temperature in terms of the distribution function which is best expressed in terms of the reduced remanence M_r/M_s for a system with randomly oriented easy axes.

$$\frac{M_r}{M_s} = 0.5 \left[1 - \int_0^{V_p} f(V) \, dV \right]$$

(3.23)

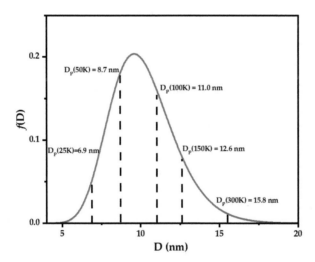

Fig. 3.5: Critical size for superparamagnetic behaviour as a function of the temperature of measurement.

This equation must be in terms of volume because it is the volume of magnetic material that is blocked that controls the remanence. The equation should in principle include the distribution of the anisotropy constant K but as stated previously it is common practice, particularly with fine particle systems, to assume a constant effective value of K particularly when the system has randomly oriented easy axes of the particles.

This simple relationship leads to an extremely useful measurement for the determination of the effective value of K because the remanent magnetisation will be at its maximum at very low temperatures and will decay with temperature producing a curve known as the temperature decay of remanence (Gittleman *et al.*, 1974). An example of the temperature decay of remanence for a system of fine particles achieved by freezing a ferrofluid in zero field down to helium temperatures is shown in Fig. 3.6. Because we are now measuring remanence, the energy barrier to reversal of each particle is simply $\Delta E = KV$.

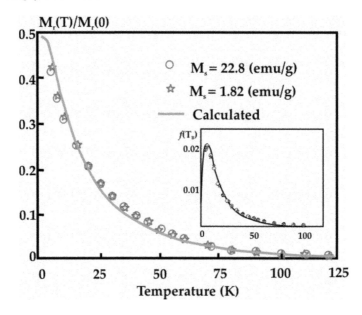

Fig. 3.6: Typical temperature decay of remanance measurement for a frozen ferrofluid with the calculated fit to the data. The inset shows the derivative of the data (O'Grady *et al.*, 1993).

However care must be taken when measuring the remanence of any fine particle system because sample shape demagnetising effects will still be there and hence a sample e.g. in the shape of a cylinder, must have an axial ratio of at least 3 and preferably 5, to ensure that the demagnetising field is insignificant. In principle, given that this is a system with randomly oriented easy axes, the remanence when the entire system consists of blocked particles should be $0.5M_s$ as discussed in Section 2.4. Occasionally this is not the case because the value of M_s will include the alignment of disordered spins at the surface of the particles which will not contribute to the remanence but do contribute to the value of M_s.

At the point where the remanence has decayed to half its initial value at very low temperatures, the size of the particles being activated is equal to that of a particle having the median volume. Note that again this is not the median diameter. A simple substitution of the value of the median volume, $< V >$, of the particles determined by transmission electron microscopy then reveals an effective value for the anisotropy constant following

$$K < V >= ln(t_m f_0) k T_B \qquad (3.24)$$

where t_m is the time of measurement which is now the time taken for the field to sweep from the saturating value to zero. It is also possible to now fit the data to a calculated curve using the effective value of K and such a calculated fit is also shown in Fig. 3.6 (O'Grady *et al.*, 1993). Differentiation of this curve then reveals the effective energy barrier distribution in the system which is shown in the inset of Fig. 3.6.

The situation regarding the coercivity is somewhat more complicated. When the coercivity is measured there is now a negative field applied to the system. The effect of this negative field will be to reverse a fraction of the blocked particles i.e. those with the lowest energy barriers. However the negative field will also induce a reversible change in the magnetisation of the system which at a temperature where all the particles are not blocked. This will include partial alignment of the superparamagnetic grains but particularly for a randomly oriented system, will also include some partial rotation of the moments of particles that are not perfectly aligned with the field. Furthermore the value of D_p at that field will not be the same as the value that occurs in zero field and the energy barrier is now given by eqn 3.10. Hence the coercivity now becomes the point at which the magnetisation of the blocked particles aligned in the original direction of the saturating field is exactly balanced by the magnetisation of the blocked particles which have reversed plus the reversible component of the magnetisation i.e.

$$\int_{V_p(H_c)}^{\infty} f(V)\,dV = \int_{V_p(0)}^{V_p(H_c)} f(V)\,dV + \int_{V_p(0)}^{0} L(\alpha)f(V)\,dV \qquad (3.25)$$

Here we have represented the reversible component via a Langevin function $L(\alpha)$ which may well be an approximation. Equation 3.25 is for the case when there is a mixture of blocked and superparamagnetic particles in the system at the temperature of measurement. However an equation of a similar form would apply to any system where there is the possibility of reversible and irreversible changes in the magnetisation at the temperature of measurement. This would include other ferromagnetic systems consisting of multi domain entities where reversible domain wall motion would be the reversible term.

Given this criterion for defining the coercivity of a magnetic system it is somewhat surprising that this parameter is generally used as the defining parameter for almost all magnetic materials. From eqn 3.25 it is clear that the coercivity is simply the limit on the integral and is therefore not an intrinsic parameter at all. For example it is clear that the coercivity is not the average switching field. A further anomaly occurs as discussed in Section 3.5, that at finite temperatures, there will exist a time dependence of the magnetisation around the value of the coercivity and hence this parameter will intrinsically depend upon the sweep rate when the measurement is made in a DC field, or the frequency of measurement when an AC technique is used. As will be discussed in Chapter 7 the magnetic recording industry which is primarily concerned with the average switching field of the grains in a medium, has for some time used a parameter known as the remanent coercivity (H_r) to characterise the materials used in discs and tapes. The measurement of H_r is discussed in Section 3.6.

3.5 Time and Frequency Effect

3.5.1 Time Dependence Effects

As discussed previously the dependence of D_p on the time or the frequency of measurement when an AC technique is used, means that key parameters vary significantly as indicated in eqn 3.12. This is because the factor 25 in eqn 3.12 applies only for a measurement time of 100 s. The general case is when a factor $ln(tf_0)$ is used. It should be emphasised that this applies to all magnetic materials whether particulate and granular or larger scale systems reversing by domain wall motion, because all systems will contain a distribution of switching fields. This in turn means that the time dependence of the magnetisation will depend not only on the time of measurement and the temperature, but also on the field sweep-rate used. The level of time dependence observed will depend on the energy barrier distribution and at what point the measurement is made within the distribution which is dependent upon the applied field.

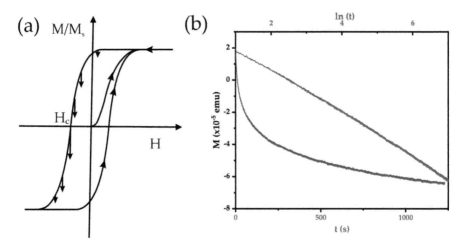

Fig. 3.7: (a) Schematic of the time dependence effects at different points in the hysteresis loop (b) Magnetisation as a function of t and $\ln(t)$ for the magnetic strip of a UK rail ticket.

There is a very simple experiment that can be done on any hysteretic magnetic material to demonstrate this effect. Any particulate magnetic material such as the magnetic strip on a bus or train ticket or a credit card (not recommended!) will show this effect quite clearly. If a small sample of the magnetic strip is cut out and measured with the field in the plane then the demagnetising field is zero. Using an instrument such as a vibrating sample magnetometer the sample is first saturated in say a positive field and the hysteresis loop measured to determine the coercivity (H_c). If the sample is now re-saturated and the field returned to a negative value in the region of H_c and held constant for a few minutes, a time dependent migration of the magnetisation towards negative saturation will be observed. The rate of migration is observed to vary with field as shown schematically by the length of the arrows in Fig. 3.7(a) and experimentally in Fig. 3.7(b). It should be noted that we expect the time dependence to be largest close to the peak in the energy barrier distribution. This arises because the energy barrier distribution closely follows the particle volume distribution and near the peak in the energy barrier distribution there are more grains that are subject to the thermal activation effect. Hence the arrows indicating the level of time dependence shows a maximum in the region of, but not exactly at, the coercivity because the coercivity includes the reversible component of the magnetisation.

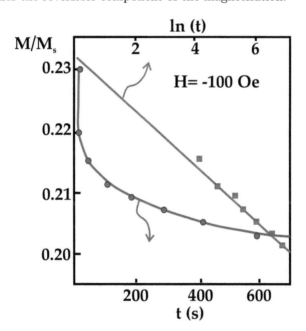

Fig. 3.8: Time dependence effects for a system consisting of randomly oriented Co particles (O'Grady *et al.*, 1981).

Figure 3.8 shows an example of the resulting time dependence which is not linear in time but also does not follow the exponential relationship that would be expected from eqn 3.4. This is because there are grains having different sizes all reversing simultaneously at different rates (O'Grady *et al.*, 1981). In principle therefore a summation of terms such as that given by eqn 3.5 would be required. However in practice it has been shown experimentally and determined theoretically, that the decay of magnetisation in systems containing a wide distribution of energy barriers, that the decay of magnetisation follows a relationship of the form (Street and Woolley, 1949; Gaunt, 1986)

$$M(t) = M(0) \pm S(H)ln(t) \tag{3.26}$$

where the parameter $S(H)$ is known as the time dependence coefficient which will depend upon the applied field and the $\pm$ sign indicates whether the magnetisation at that field is increasing or decreasing. It should be noted that such time dependence effects can be observed on a hysteresis loop taken in the reversal regions on the loops but can also be observed on the magnetising part of the measurement from an initially demagnetised state. In the example shown in Fig. 3.8 the decay in the magnetisation is over 10% in 10 minutes. However as this sample is a frozen ferrofluid containing Co particles of diameter 8 nmn randomly oriented, the switching region is very wide. For other materials such as a metallic thin film where the grains are strongly exchange coupled, the hysteresis loop is highly square. In such samples we have observed changes from $+0.5M_s$ to $-0.5M_s$ in less than two minutes.

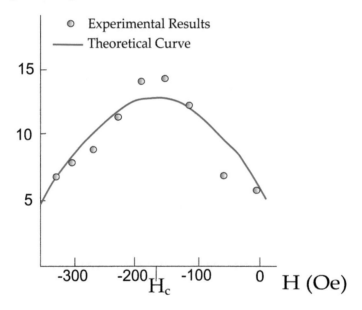

Fig. 3.9: (a) Variation of the magnetic viscosity coefficient as a function of the applied field (O'Grady *et al.*, 1981). The theoretical curve was calculated by R. W. Chantrell and is based on an integral across the particle size distribution.

Figure 3.9 shows the value of the time dependence coefficient determined from the slope of the data in Fig. 3.7(b) as a function of the applied field. As can be seen the parameter $S(H)$ has a very similar form to that of the switching field distribution. However the peak in the curve does not occur at the coercivity because of the reversible component of magnetisation. It is only blocked particles reversing over their energy barriers that give rise to the time dependence. Hence the peak occurs at the peak in the switching field distribution discussed in Section 3.6. The peak will occur at the peak in the energy barrier distribution which can either be determined from a temperature decay of remanence measurement discussed in Section 3.4, but is correctly determined from a field dependence of the switching field of the individual grains. The line in Fig. 3.9 was calculated from a theoretical analysis by R. W. Chantrell (O'Grady *et al.*, 1981) which gave

$$S(H) = S_{max} exp - \left[ln \left(\frac{H_K - H_c}{H_K - H} \right)^2 \frac{2}{9\sigma^2} \right]$$ (3.27)

In an important but little studied work, Gaunt (1986) investigated the parameters that control the value of $S(H)$ and showed that

$$S(H) = 2kTM_s f(\Delta E)|_{\Delta E_c(H)}$$ (3.28)

where $\Delta E_c(H)$ is the critical energy barrier being activated in the field H. Assuming K to be constant

$$< \Delta E_c(H) > = K < V > (1 - H/H_K)^2$$ (3.29)

The first 4 parameters in eqn 3.28 are constant so $S(H)$ is directly proportional to the value of the energy barrier evaluated at the critical energy barrier $(\Delta E_c(H))$ being activated by the applied field H. This variation is key to understanding time dependence in real materials.

The first observation of the logarithmic nature of the time dependence was made by Street and Woolley in 1949 who studied the effect in Alnico permanent magnets. However there have been many such measurements subsequently (e.g. Ford, 1982; Uren *et al.*, 1988; O'Grady and *et al.*, 1981) all of which have found similar behaviour.

In the 1990s there was a great deal of controversy relating to materials such as spin glasses and certain types of thin films where time dependence was observed that was not linear with $ln(t)$. In an attempt to explain this data it was suggested that different forms of distribution function could be responsible (Aharoni, 1985) but in a defining work, el-Hilo *et al.* (1992a) showed that the non-linearity could be fully explained by integration across an energy barrier distribution which was able to explain the origin of some of the apparently anomalous data when plotted on an $ln(t)$ axis. The variation of the magnetisation with time was observed to curve downwards i.e. accelerating for some field values, whereas for others it was observed to slow down. At certain fields the plot of M versus $ln(t)$ was observed to follow an S shape.

However it was noticed that materials exhibiting non-$ln(t)$ behaviour were always those that exhibited a very narrow switching field distribution as exhibited by a highly square hysteresis loop. These included metallic spin glasses, exchange coupled metallic thin films such as rare-earth-transition-metal alloys with perpendicular anisotropy proposed at the time for magneto-optic recording systems, and NdFeB permanent magnets.

Referring to the Gaunt equation (eqn 3.28) the $ln(t)$ variation defined by the time dependence coefficient requires that the value of $f(\Delta E)$ is constant or at least approximately so for the grain sizes being activated by the field H and thermal energy. This is only true when $f(\Delta E)$ is relatively wide, based on the summation of exponential

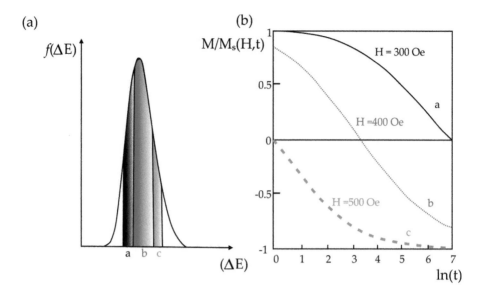

Fig. 3.10: (a) Schematic diagram of the energy barrier distribution for a system with a narrow distribution ($\sigma = 0.1$) and (b) Variation of the magnetisation with t and $ln(t)$ for a system with a narrow distribution of energy barriers (el-Hilo *et al.*, 1992a).

decays given by eqn 3.3. Consider now Fig. 3.10(a) which shows a distribution $f(\Delta E)$ where $\sigma=0.1$. This unusual set of observations could be readily explained in terms of eqn 3.28. The dependence of the degree of linearity in the time dependence depends upon the value of S which in turn depends upon the value of $f(\Delta E)$. Hence if the value of $\Delta E_c(H)$ lies at low values to the left of the peak in the energy barrier distribution (region a), then the time dependence will accelerate with time as slightly higher energy barriers are activated. Similarly if the value of ΔE_c lies to the right of the peak (region c), then the rate of change will be observed to decrease. The anomalous S shaped curves occur when the measurements start very close to the peak in the energy barrier distribution (region b), but as time progresses those energy barriers being activated lie to the right of the peak and hence decelerate. These positions for the required narrow distribution are shown in Fig. 3.10(a) and the resulting form of M vs $ln(t)$ is in Fig. 3.10(b) which is the original calculation of el-Hilo *et al.* (1992a).

3.5.2   Frequency Effects

Whilst the time dependence of magnetisation invariably leads to a frequency dependence of parameters such as the coercivity, in general most studies undertaken on this subject have looked at the variation in the coercivity as a function of the field sweep rate. This has occurred because most measurements on particulate and granular systems are generally undertaken using instruments such as a vibrating sample magnetometer or an alternating gradient force magnetometer. Hence the analysis is generally done in terms of a field sweep rate $dH/dt = R$. In the 1990s this topic was of considerable concern because of the effects particularly on recording media, with the very large difference in effective sweep rate between the writing process that was at that time generally at 10s of MHz and the storage of the information that was required to be stable for a period of around 10 years. This required these effects to be understood both for the design of the write head and the design of the media. At that time the information was stored in bits in the plane of the tape or thin film disk such that a significant demagnetising field was present both from the bit acting on itself but also from neighbouring bits that would also generate a further demagnetising field. A number of theoretical analyses of this effect were undertaken notably by Sharrock (1990) and Doyle *et al.* (1993) but the final detailed analysis based on the full relaxation time formula of Brown determined that the description of the sweep rate dependence of the coercivity was best represented by a ln(R) expansion of the form (el-Hilo *et al.*, 1992b)

$$H_c(R) = H_K - H_K \left(\frac{ln\left(f_0 H_K / 2\alpha_m R_0\right)}{\alpha_m} \right)^{1/2} \left[1 - \frac{ln\left(R/R_0\right)}{ln\left(f_0 H_K / 2\alpha_m R_0\right)} \right]^{1/2} \quad (3.30)$$

where $\alpha_m = K < V > /kT$, R was the sweep rate and R_0 was the initial value of the sweep rate. It was also shown that only taking this expansion to two terms resulted in a deviation of <2%. Figure 3.11 shows the variation of coercivity with sweep rate for a sample of metal particle tape which at that time was the most advanced tape media available. From this data it can clearly be seen that the material broadly followed the trend predicted by eqn 3.30.

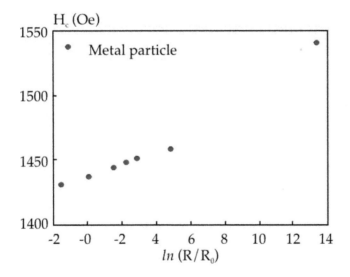

Fig. 3.11: Measured sweep rate dependence of the coercivity for metal particles. Reprinted from Journal of Magnetism and Magnetic Materials., 120 , de Witte et al., Sweep rate measurements of coercivity in particulate recording media, 184, Copyright (1993), with permission from Elsevier.

3.6   Switching Field Distributions

From a purely scientific point of view and indeed for applications particularly in information storage systems, it is essential to have a detailed knowledge of the distribution of switching fields within a given material whether particulate or a granular thin film. This topic first came to the forefront during the era when magnetic tape recording was dominant in information storage and the first attempt to characterise the switching field distribution was made by Eberhard Koester of BASF (1984). The technique relies on the analysis of a standard hysteresis loop as shown in Fig. 3.12. As shown in the diagram, a straight line is fitted to the measured hysteresis loop through the region of the coercivity and extrapolated to the point of remanence in a positive field. The distance from the extrapolated line to the $H = 0$ axis is then denoted as a fraction (S^*) of the coercivity given by $S^* H_c$. The parameter $1\text{-}S^*$ is then taken as a measurement of the width of the switching field distribution.

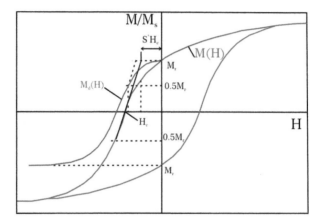

Fig. 3.12: Schematic of the Koester method used to measure the switching field distribution.

Of course this technique suffers from the deficiency that the hysteresis loop contains both reversible and irreversible components of the magnetisation so the slope of the curve through the coercivity is not a true reflection of the switching of the grains exclusively. None-the-less this technique was used for many years from its first description almost to the end of the 20th century. To add a measure of confusion the hard disk industry also used this technique but rather than quoting the value of $1\text{-}S^*$ as the key parameter, they preferred to use the parameter $S^* H_c$. For the case of thin film media used in hard drives this was perhaps more representative because the level of reversible magnetisation was significantly lower than was the case for tape systems. However it remains that the Koester technique is fundamentally flawed because a measurement directly from a hysteresis loop will always contain elements of reversible magnetisation.

As we have seen in Section 3.4 it is possible to measure a distribution of energy barriers via the temperature decay of remanence. However for real systems it is generally required to measure the distribution of switching fields, which is in effect a measure of the distribution of the energy barriers, at a constant temperature. To do this requires a technique that removes the reversible components of the magnetisation which is relatively simple by measuring a field dependent remanence curve.

There are two forms of field dependent remanence curves known as the isothermal remanence magnetisation (IRM) curve ($M_r(H)$) where the acquisition of remanence is measured from a demagnetised state. The alternative is to measure the removal of remanence from a saturated state which is known as the DC demagnetisation (DCD) curve denoted $M_d(H)$.

However as was the case for the temperature decay of remanence, because this measurement is made in zero applied field it is necessary to take account of any sample shape demagnetising effects. Hence for a powder or frozen ferrofluid sample it is necessary to use a sample holder which has an axial ratio of at least 3 and preferably 5. However for a thin film sample or e.g. for a coating such as a magnetic tape the magnetisation lies in the plane of the film and hence the sample shape demagnetising factor will be zero. In fact for some powder samples it can be advantageous to produce a sample in the shape of a coating by binding the particles in a resin onto some substrate rather than create a 3-dimensional structure.

The IRM curve is measured from the initially demagnetised state by taking the sample through successive minor loops. When undertaking this measurement it is often advisable to ensure that the sample is fully demagnetised before starting by exposure to a large AC field in a coil from which the sample is very gradually withdrawn.

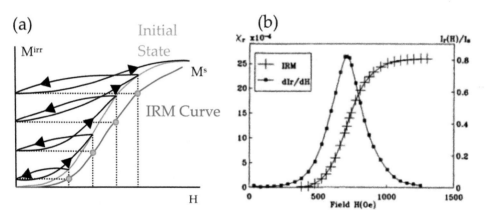

Fig. 3.13: (a) Schematic of an IRM measurement and (b)measured IRM curve and its differential. Reprinted from Elsevier books, North-Holland Delta Series O'Grady and Chantrell, In Magnetic Properties of Fine Particles, 93-102, Copyright (1992), with permission from Elsevier.

The measurement procedure is then shown schematically in Fig. 3.13(a) and in Fig. 3.13(b) is shown the resulting remanence curve and its differential. The differential of the IRM curve is technically a measurement of susceptibility because it is the slope of a curve of magnetisation against field. In this case the slope is the change in the irreversible magnetisation and is generally denoted χ_r^{irr}. The superscript denotes that this is the irreversible susceptibility and the subscript indicates that it is taken from an IRM curve.

In a similar manner a remanence curve can be obtained by progressive demagnetisation of a previously magnetised sample. The measurement process for the DCD curve is shown in Fig. 3.14(a) and in Fig. 3.14(b) a set of experimental data and the resulting differential of the DCD curve is shown. Again the differential represents a value of the irreversible susceptibility but taken from a DCD curve and is generally denoted χ_d^{irr}.

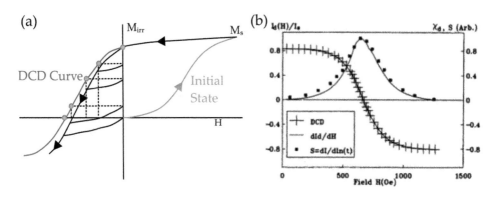

Fig. 3.14: (a) Schematic of a DCD measurement and (b) measured DCD curve and its differential. Reprinted from Elsevier books, North-Holland Delta Series O'Grady and Chantrell, In Magnetic Properties of Fine Particles, 93-102, Copyright (1992), with permission from Elsevier.

To obtain the switching field distribution (SFD) from either the χ_r^{irr} or χ_d^{irr} data it is necessary to normalise the χ^{irr} curves to unity. Hence

$$SFD = \frac{\chi_{irr}}{\int_0^\infty \chi_{irr}(H)dH} \tag{3.31}$$

where the integral generates the necessary normalisation factor.

Because of the shape of both the IRM and DCD curves, but particularly the latter, fitting a curve to $\chi^{irr}(H)$ can be somewhat challenging. In particular the use of graphic packages and certainly the use of cubic spline fitting techniques, generally produces a poor fit to the data. In our laboratories some years ago we generated our own software to fit these sigmoidal curves which was based on a summation of quadratic fits to a part of the curve typically consisting of 5 points and then removing the first point and adding the 6 point etc. and taking a slope at the mid-point of each 5 point quadratic fit. This technique proved very successful in fitting the experimental measurements and producing a smooth fit to the data as shown in Fig. 3.13(b) and 3.14(b).

In principle the data obtained from an IRM and a DCD curve should be identical differing only in a factor 2 because the IRM curve goes from zero to M_s whereas

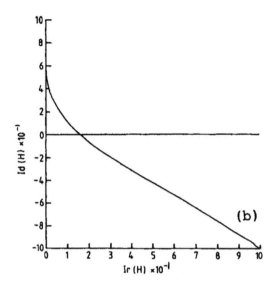

Fig. 3.15: Example of a Henkel plot. Reprinted from Elsevier books, North-Holland Delta Series O'Grady and Chantrell, In Magnetic Properties of Fine Particles, 93-102, Copyright (1992), with permission from Elsevier.

the DCD curve goes from $+M_s$ to $-M_s$. This fact was first noted by the late E. P. Wohlfarth who wrote

$$M_d = 1 - 2M_r \tag{3.32}$$

Equation 3.32 is known as the Wohlfarth relation (Wohlfarth, 1958) and is applicable when there are no interactions within the system. Of course this is almost never the case. Henkel (1964) produced plots of the relationship shown in eqn 3.32 and described deviations from the expected straight line as being due to dipolar and possibly exchange interactions in the material. An example of such a Henkel plot is shown in Fig. 3.15 for a system with only dipolar interactions and hence the data lies below the line (Chantrell and O'Grady, 1992). The issue of dipole-dipole interactions in particulate and granular materials will be discussed in more detail in Section 3.8.

3.7 The Fluctuation Field and Activation Volume

In the study of thermal activation processes in magnetic materials there is an inherent difficulty in having a mathematical formulation that couples the thermal energy in the material to the magnetic moment. Néel (1949b) postulated that this coupling of thermal energy to the moment could be represented by a fluctuating magnetic field (H_f). He related the fluctuation field to the value of the thermal energy kT in the following way

$$H_f = \frac{kT}{Q} \tag{3.33}$$

where Q was an undefined parameter. Wohlfarth (1983) arguing on dimensional grounds pointed out that the undefined parameter Q in eqn 3.33 must have the dimensions of a magnetic moment and re-wrote the equation as

$$H_f = \frac{kT}{V_{act}M_s} \qquad (3.34)$$

Wohlfarth described the parameter V_{act} as an activation volume which was that volume of the material that was activated by the thermal energy. Néel (1949b) also pointed out that the fluctuation field given by eqns 3.33 or 3.34 could be measured experimentally as the ratio between the logarithmic time dependence of the magnetisation discussed in Section 3.5, divided by the value of the irreversible susceptibility described in Section 3.6 where the irreversible susceptibility must be measured in the same quadrant of the hysteresis process as the measurement of the time dependence.

$$H_f(H) = \frac{S(H)}{\chi_{irr}(H)} \qquad (3.35)$$

Of course there is a further requirement that the magnetic viscosity ($S(H)$) must be determined for systems where the variation of the magnetisation with time is linear with $ln(t)$. As we have seen in Section 3.5 this is not always the case. Furthermore the field dependence of both S and χ^{irr} in eqn 3.35 implies that the fluctuation field (H_f) will also be field dependent. This hypothesis seems perfectly reasonable if one considers a system of discrete magnetic nanoparticles where both the irreversible susceptibility and the time dependence will vary with the particle size. Assuming the magnetisation reversal process to be coherent rotation in particles of say 10 nm, then the activation volume will vary with the particle size and hence so will the fluctuation field. A number of studies were undertaken in our laboratories on dilute fine particle systems created by simply freezing a ferrofluid, where the particle size is typically of the order of 10 nm and to a first approximation it was confirmed that the activation volume defined by Wohlfarth correlated with the particle size in the system (de Witte *et al.*, 1990).

A more surprising result was obtained when we undertook studies of the elongated particles then used in magnetic recording tape and specifically video tape (de Witte *et al.*, 1993). Such particles are typically elongated and made of doped iron oxide with an axial ratio lying in the range 5 to 10. For these systems it was found that the value of the activation volume measured near the coercivity, corresponded with a fraction of the particle volume related to its elongation i.e. a particle with say an elongation of a factor 7, would have an activation volume equivalent to one seventh of the volume of the particle. This is a somewhat astonishing result but it indicates that the incoherent reversal mechanism which was known to occur in such particles, closely corresponded to the chain of spheres model (see Section 2.5) described by Jacobs and Bean (1955) for the incoherent reversal mechanism which was occurring.

A further surprising result was obtained on studies of the then longitudinal thin film media used in hard drives. Here it was found that the activation volume of reversal in such thin films where significant exchange coupling between the grains was known to exist, was significantly larger than the volume of an individual grain implying that a number of grains were reversing as a single entity due to the exchange coupling. Some attempt was made to vary the degree of intergranular exchange coupling in such films as described in Section 2.6 and again a correlation was found between the activation volume and the anticipated degree of reduced intergranular coupling (O'Grady *et al.*, 1998).

Hence the use of the activation volume even in the simplest form described by eqn 3.34, was found to reflect in some detail the nature of the reversal process ongoing in the material. The verification of the chain of spheres type reversal was quite conclusive but the correlation of the activation volume to the degree of exchange coupling in a longitudinal thin film medium also correlated to the level of electronic noise generated in recording signals. Hence the activation volume became much more than a concept of academic curiosity as it related closely to observed engineering phenomenon in the materials.

We must now consider the measurement of the activation volume when the variation of magnetisation with time is not linear with $ln(t)$. As we have seen this is the case where the width of the switching field distribution is narrow. This problem was first addressed by the group of McCormick and Street of the University of Western Australia (1992) who studied NdFeB magnets. Here the loop is nucleation controlled giving a square loop (Estrin *et al.*, 1989). They considered an equation of state formulation and wrote

$$\chi^{irr} = \frac{\partial M_{irr}}{\partial H}\bigg|_{\dot{M}_{irr}} \tag{3.36}$$

$$H_f = \frac{\partial H}{\partial \dot{M}_{irr}}\bigg|_{M_{irr}} \tag{3.37}$$

This formulation suffers from the obvious difficulties of measuring the various parameters at a constant value of the rate of change of the irreversible magnetisation but none-the-less was found to be successful if suitable computer control over the vibrating sample magnetometer or other instrument were achieved.

However el-Hilo *et al.* (2002) modified this equation of state to obtain formulae that were much more experimentally accessible.

$$H_f = \frac{\Delta H}{\Delta ln(\dot{M})}\bigg|_{M_{irr}} \tag{3.38}$$

where for the case when $\Delta ln(t) >> \Delta ln(S)$ values of H_f can be obtained readily from

$$H_f = \frac{-\Delta H}{\Delta ln(t)}\bigg|_{M_{irr}} \tag{3.39}$$

$$H_f = \frac{kT}{\partial \Delta E(H)/\partial H|_{\Delta E = \Delta E_c}} \tag{3.40}$$

or

$$H_f = \left|\frac{\Delta H}{ln(t_2/t_1)}\right|_{M_{irr}} \tag{3.41}$$

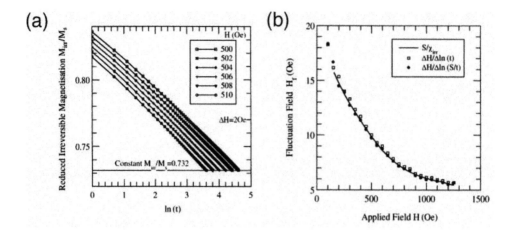

Fig. 3.16: (a) Example of time dependence data (a) used to calculate H_f and (b) showing the variation of H_f with the applied field. Reprinted from Journal of Magnetism and Magnetic Materials, 248, el-Hilo *et al.*, Fluctuation fields and reversal mechanisms in granular magnetic systems, 360, Copyright (2002), with permission from Elsevier.

This formalism leads to the measurement of H_f being determined from a simple waiting time experiment where the time for the magnetisation to decay to a constant value in a given reverse applied field is measured. Hence not only can H_f be determined e.g. at the coercivity but it can also be determined as a function of field. An example of such data showing the decay to a constant value of M is shown in Fig. 3.16(a) calculated theoretically for a system having standard deviations of volume of a lognormal distribution function of 0.52 and a standard deviation of the anisotropy constant equal to 0.1 and the variation of H_f with the applied field in the system is in Fig 3.16(b). The requirement that the waiting time measurement is made at constant M_{irr} can be achieved for systems with a highly square loop as the change in M is entirely irreversible. Importantly this applies to all current recording media which are based on hcp Co-alloy grains having almost perfect c-axis alignment perpendicular to the plane of the film.

A full discussion of the determination of the fluctuation field from equation of state formulations is provided in el-Hilo *et al.* (2002). The determination of the fluctuation field and the activation volume for non-linear systems became more important with the development of perpendicular magnetic recording where due to the strong orientation of the grains in the thin films grown, and to some degree the effect of exchange coupling between the grains, resulted in highly square hysteresis loops. The data for these materials also has to be considered in terms of the now very strong demagnetising field perpendicular to the plane of the films which in effect will be equal to $-4\pi M_s$. The first study of such materials was undertaken by Dutson *et al.* (2003). Again a correlation was found between the activation volume and the degree of intergranular exchange coupling engineered in the films which were produced by Seagate Media Research in Fremont CA.

At the time of writing all hard drive storage media consists of films with very strong perpendicular orientation of the easy axes of the grains and hence for such modern systems including those used for heat assisted magnetic recording (HAMR), methods such as the waiting time method for the determination of H_f and V_{act} have to be employed. Also with the rapidly advancing development of spin transfer torque MRAM (STT-MRAM) discussed in Chapter 8 once again the orientation of the easy axis of the grains in the patterned films is perpendicular to the plane of the film and hence such techniques have to be used because the time dependence of the magnetisation will be non-linear in $ln(t)$.

It should also be noted that measurements on such films are best made with a focused Magneto Optic Kerr Ellipsometer (MOKE) magnetisation measurement system. In particular the use of a focused MOKE also allows for small areas of a thin film to be measured allowing some estimation of the uniformity of small areas of the film to be obtained. Similar data should be possible by determining the value of the resistivity of individual elements in patterned films and then integrating over a number of such pixels. The use of the fluctuation field technique in MRAM elements has been undertaken by Carpenter *et al.* of the IMEC Institute in Leuven, Belgium (private communication, 2021).

3.8 Intergranular Interactions

3.8.1 Dipole-Dipole Effects

In Section 2.6 we discussed the occurrence of intergranular exchange coupling in thin film materials which can only occur when the material of the thin film is electrically conducting. This allows for indirect exchange coupling to occur via the Ruderman-Kittel-Kasuya-Yosida (RKKY) phenomenon. Such interactions are very strong and can lead to cooperative reversal over many grains resulting in an effective domain structure in the resulting film assuming that it is dense. As we will see in Chapter 7 a great deal of effort has gone into limiting the effect of RKKY interactions in certain classes of metallic thin films notably for information storage, so that the size of the resulting bits when written to the medium can be reduced significantly.

However in all particulate and granular materials there will always be dipole-dipole interactions between the grains in the film regardless of whether or not they are conducting or insulating. In metallic thin films in particular, this can lead to extremely complex behaviour but in the first instance we will simply consider the dipole-dipole interactions which are ubiquitous in all particulate and granular systems.

In the case of the dipole-dipole interaction we can first consider the simplest case where two identical magnetic particles come into close proximity in which case dipole-dipole interactions are magnetising in character. There is often a misconception that dipole-dipole interactions are always demagnetising. It is simple to show that if the particles are free to move or the material of which they are composed has limited anisotropy, that the particles will tend to align with their magnetic moments parallel to each other. Such interactions can be observed in materials such as ferrofluids discussed in Chapter 6 where the particles are suspended in a liquid and are free to move. In very dilute systems such as those used to undertake studies using TEM images, chains of particles have been observed to occur particularly if the sample was allowed to dry

(a)   (b)

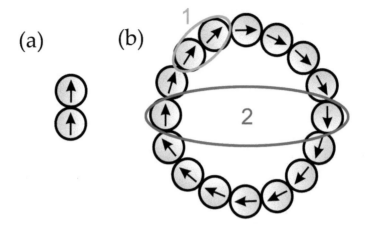

Fig. 3.17: Schematic configuration for (a) two isolated nanoparticles and (b) a ring of nanoparticles.

in the presence of even a very weak magnetic field and on occasions in zero field. Rings of particles have been observed and in Fig. 3.17 the configuration for (a) two isolated identical particles and (b) for a ring of particles is shown schematically.

In principle the energy of two interacting magnetic particles each consisting of a single magnetic domain is given by

$$E_{d-d} = \frac{m_1 m_2}{r_{1-2}^3} \ (cgs) \tag{3.42}$$

where m_1 and m_2 are the magnetic moments of the dipoles and r_{1-2} the centre to centre separation. However the assumption used to derive eqn 3.42 is that the separation of the particles r is much greater than the size of the dipoles themselves. In most magnetic systems and certainly in thin films, this is not the case and it can be shown that the variation of the dipole-dipole energy is much closer to an r^2 variation. This is certainly the case in relatively dense thin films such as those commonly studied and is also the case in materials such as ferrofluids and powders of materials where the particles are either free to move or in relatively close proximity to each other.

If we now consider the ring of particles shown in Fig. 3.17(b) the orientation of the moments in the particles is fairly obvious in that the direction of the moments assuming the particles are of low anisotropy or free to move, will follow the shape of the ring. If we now consider the two particles contained within the box 1 in the figure or indeed any other two particles which are adjacent to each other, then it is clear that the interaction is predominately magnetising in nature in that the direction of the moments in the particles reinforce each other. However if we now consider the two particles in box 2 on opposite sides of the ring or indeed any two particles which lie opposite each other, then it is clear that the moments on those particles will be in opposite directions and hence will be demagnetising in nature. As one looks at the orientation of the moments around the rings relative to one moment, the nature of the interactions shift from magnetising in the short distance to fully demagnetising at maximum separation. Hence it is a misconception to say that dipole-dipole interactions are always demagnetising in character although we will see that overall, exclusively dipole-dipole interactions are demagnetising as shown in Fig. 3.15.

3.8.2 RKKY Exchange Interactions

When the situation in a metallic thin film is considered then the presence of the RKKY interaction which is generally magnetising in nature leads to a highly complex many body problem. Given the size of single domain particles in either powder, liquid or thin film form this is a many body problem running to billions if not trillions of particles per mm^2. The only way that the effect of interactions on this scale can be addressed is by large scale computer modelling usually involving the use of cyclic boundary conditions so that the cell itself which may contain several hundred or preferably several thousand particles, becomes representative of the whole sample of the material.

In the case of discrete particles the development of such large scale modelling was pioneered by our colleague R. W. Chantrell whose first model (Menear *et al.*, 1984) was a study of ferrofluids where the system was more complex because of the ability of the particles to move. In the case of granular thin films used for hard disk media the development of the early models was pioneered by a longstanding collaboration by Zhu and Bertram (1989). Because of the scale of the hard disk drive industry, over the last 20 years or more, there have been a large number of computer models that have been developed looking at a specific set of circumstances or with specific assumptions. For example the original work of Zhu and Bertram (1989) where the particles were placed on an hexagonal lattice was shown to lead to artifacts in the results because the particles in a genuine thin film are positioned randomly on the substrate.

In this text it is not possible and perhaps not desirable to review even some of the models quantitatively and the interested reader is referred to a number of suitable references and book chapters which address the subject in some detail. However there are a number of works which demonstrate the underlying physics of the behaviour of such complex interacting systems.

In Section 3.6 we described the methods for the correct measurement of the switching field distribution via the measurement of remanence curves and describe how the difference between the IRM and DCD curves leads to the well known Wohlfarth relationship and the use of Henkel curves such as that shown in Fig. 3.15. Henkel curves are useful only for the case where there is a system having only dipole-dipole interactions. On the macroscopic scale given that dipole-dipole interactions are generally demagnetising, the measurements will differ from the predicted straight line lying beneath the line indicating that the interactions are reducing the level of remanence in the system. However when one considers metallic thin films the nature of the experimental data becomes far more complex with deviations both above and below the line prohibiting ease of interpretation.

In 1989 Kelly *et al.* showed that the Wohlfarth relationship could be reformulated to look at the deviations (ΔM) from linearity in the Henkel plots explicitly.

$$\Delta M = M_d(1 - 2M_r) \tag{3.43}$$

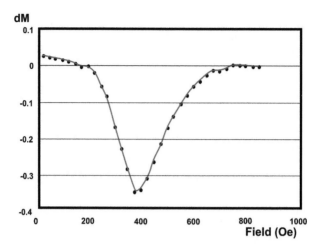

Fig. 3.18: ΔM plot for CrO_2 data tape.

Figure 3.18 shows such a ΔM curve for a sample of data tape made around the same time as the work of Kelly *et al.* (1989) where the tape consisted of particles of CrO_2 which was a standard storage format at the time. This result shows that the interactions in a tape made from discrete particles held in a polymer binder are overall demagnetising in nature as expected.

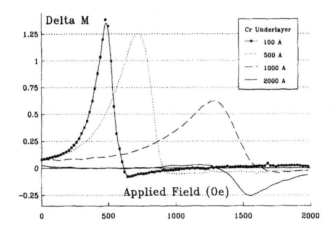

Fig. 3.19: Delta M plots as a function of the Cr underlayer thickness. Reprinted from Journal of Magnetism and Magnetic Materials, 95, Mayo *et al.*, Magnetic measurement of interaction effects in CoNiCr and CoPtCr thin film media, 109, Copyright (1991), with permission from Elsevier.

However subsequently Mayo *et al.* (1991) produced a defining study on a set of idealised longitudinal recording media consisting of an alloy of CoPtCr thin films grown on an underlayer of Cr where increasing thickness of the Cr underlayer resulted in spatial segregations of the magnetic alloy grains in the film. In this way the effect of the RKKY interactions could be controlled and progressively reduced as the thickness of the Cr layer was increased. Figure 3.19 shows the variation of ΔM for the alloy films grown on the different thicknesses of Cr (Mayo *et al.*, 1991). As can be seen as the thickness of the Cr layer is increased the gradient of the ΔM curve reduced until finally for the thickest layer (200 nm) ΔM becomes negative indicating that dipole-dipole interactions are dominant over the previously dominant RKKY effects.

In addition, in the same study it was found that there was a markened broadening of the χ^{irr} curves and also a displacement between the χ^{irr} curves for the IRM curve and the DCD curve also shifted laterally along the field axis showing in detail the effect of the interactions. Importantly the industrial collaborators on the work from IBM were able to correlate the behaviour of the ΔM curves with the observed media noise in a hard drive. They found that the gradient of the ΔM curve as it reverses from positive to negative values correlated to the integrated medium noise (Mayo *et al.*, 1991).

Zhu and Bertram (1991) were able to use their model to vary the strength of the exchange interactions between the grains and produced a remarkable correlation between our measurements and predictions from their model as shown in Fig. 3.20. This validated the ΔM technique as the only experimental technique available to quantify the interactions in a real metallic thin film system.

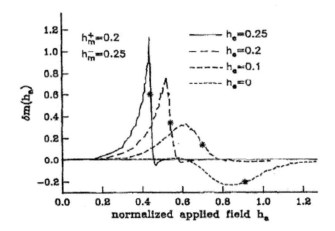

Fig. 3.20: Calculated ΔM plots by Zhu and Bertram. Reprinted from Zhu and Bertram J. Appl. Phys. 69 4709 (1991), with the permission of AIP Publishing.

Whilst the use of remanence curves and the ΔM formalism were found to be the best available technique for the study of dipolar and exchange interactions, care must be taken with these measurements as they are subject to thermal effects. By definition the measurement of a remanence curve must involve traversing part of the switching field distribution and as the field used varies from point to point, then the measurement time in eqn 3.24 is not constant. However in recent times the use of high voltage and hence high sweep-rate power supplies for resistive magnets have limited this effect. For superconducting magnet base magnetometers this remains an issue.

3.9   Susceptibility Effects

In the 1970s and 1980s there was a great deal of controversy about the behaviour of a class of materials known as spin glasses. As the name implies these were disordered alloys generally of a magnetic material such as cobalt in a conducting matrix such as copper. At the time the exact nature of the structure of these materials was not fully known but it was postulated that because the carrier matrix was a conductor, that exchange interactions could occur between individual atoms or clusters of atoms of the magnetic material. The standard measurement that was undertaken to look at the spin glass state was to cool the sample in zero field and then measure the variation of the initial susceptibility of the hysteresis loop as the temperature was increased. This produced a characteristic peak in the susceptibility usually at a temperature below that of liquid nitrogen. A wide range of models were proposed, mainly analytical in nature because large scale computer modelling was not available, to explain these effects. Other measurements of e.g. the thermo-remanence where the resulting remanence was measured as a function of the cooling temperature was taken and a vast endeavour with the publication of many papers resulting. For a summary of spin glass effects see Tholence and Tournier (1977).

The situation became somewhat turbulent when it was pointed out by Wohlfarth (1979) that a simple model of a fine particle system containing a distribution of particle sizes would also exhibit a similar peak in a susceptibility due to the unblocking of moments as the temperature was raised. The basis of the Wohlfarth explanation was that at very low temperatures all the magnetic particles within the alloy would be blocked according to the theory of superparamagnetism described in Section 3.2. Of course in a low field of typically 10 Oe used for the measurement, only a small fraction of the distribution would have the magnetic moments aligned. However as the temperature was raised the number of moments that became unblocked would increase but as the temperature was raised further particularly if the particles were very small, disorder would cause a decrease in the measured susceptibility hence giving rise to a peak.

In the subsequent years the controversy continued and as far as is known was never finally resolved to everyone's satisfaction. However in two papers el-Hilo *et al.* (1992c and 1992d) undertook a full analysis of these effects for a particulate system investigating the concentration dependence of the peak and the field dependence of the peak which was also observed. These works showed that a detailed theory of the behaviour of a fine particle or granular system it was possible to explain the behaviour of a fine particle system which was clearly in many respects analogous to that which was observed in a spin glass. However the adherence to a spin glass being a separate class of material never fully embraced the simple thermal activation model and furthermore the works of el-Hilo *et al.* did not include the possibility of exchange interactions as the theory and the measurements were based on a fine particle system consisting of a frozen ferrofluid.

Eventually the enormous interest in spin glass phenomena faded from prominence around the turn of the millennium.

3.10   Minor Loop Phenomena

As we have seen particulate and granular magnetic systems exhibit thermal activation where e.g., these effects are most manifest in the second and third quadrants of the hysteresis loop following saturation in say a positive field. These effects give rise to some rather unusual phenomenon. Consider now Fig. 3.21 which shows schematically the effect when such a system is first saturated and then the magnetisation is taken to a reverse field which is somewhat lower than the coercivity (Lewis *et al.*, 1993). The field is then swept back to $H = 0$ and subsequently returned to the same negative field as used previously and cycled between these two field points for a number of times. It is observed that the resulting minor hysteresis loops slowly decay towards $M = -M_s$ with the cycling of the loops. Néel (1959) first considered this phenomenon and decided that the fact that the magnetisation changed with field cycling must in some way be indicative of the interactions in the system assuming they were purely dipolar in nature. He termed this effect "reptation". To the best of our knowledge there is no equivalent word in English as Néel was writing in French. He then proposed that there was a demagnetising interaction field which he called the reptation field that was responsible for this effect. Unfortunately in his work he did not consider the effects of thermal activation which are close to a maximum value near the coercivity. The phenomenon attracted little attention and was merely noted as a further peculiarity in the behaviour of particulate systems.

Subsequently using the universal rate equation, el-Hilo *et al.* (1993) showed that the variation in the value of the magnetisation at the field close to the coercivity which was observed, could be predicted without the requirement to invent a reptation field. The effect was attributed to the integrated time dependence during the field sweep from zero field to the field of choice to which the measurement was taken. This was supported by a further paper from our laboratory published in the same issue of the journal indicating that this is what was indeed the case (Lewis *et al.*, 1993).

However subsequently a further study from our laboratory investigated the area of the minor loops as a function of the field applied in the reptation measurement and found that the magnetic viscosity alone would not account for the area generated in the minor loop (O'Grady and Greaves, 1995).

There is a further and as yet unexplained anomaly associated with these minor loop phenomena which is why they are included here. If one considers the schematic diagram shown in Fig. 3.21 then on consideration of the reptation phenomena there is clearly an anomaly in our understanding of these effects. We have used the schematic diagram to show the formation of minor loops in the third quadrant of a standard hysteresis loop. Figures such as these are commonly included in standard texts in magnetism when hysteresis is discussed and not just the hysteresis of particulate and granulate materials e.g. such a diagram is included in the well known text by B. D. Cullity and Graham (2009) 2nd edition on page 19. A careful examination of the

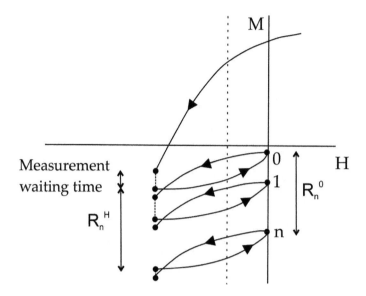

Fig. 3.21: Schematic representation of the reptation measurement (Lewis *et al.*, 1993).

figure where we have included an arbitrary dotted line parallel to the magnetisation axis and then consider the processes which have occurred then another as yet unknown phenomenon must be occurring.

Following the schematic the sample is taken from saturation to remanence and then swept out to a field greater than its coercivity. The field is then returned to zero and is again swept out to the same field where on this second loop the magnetisation has increased to above the recoil value as it was returned from the field of choice. However the magnetic field was never in a positive orientation as the measurement was terminated at the $H = 0$ point. Somehow during this process the magnetisation was observed to increase thus creating the minor loop. Thermodynamically this is equivalent to saying that the order increased in a positive sense despite the potential applied being negative. In simple terms this implies that heat flowed from a cooler body to a hotter one without energy being put in!

These relatively minor effects are included here to show that despite the enormous amount of work done on particulate and granular magnetism over perhaps almost a hundred years of endeavour, there are still effects that occur which are either unexplained or otherwise anomalous.

4
Exchange Bias

4.1 Discovery and Origin of Exchange Bias

Exchange bias was discovered accidentally in 1956 by Meiklejohn and Bean when studying the coercivity of Co fine particles prepared by electro-deposition of cobalt nanoparticles into mercury. The cobalt particles were removed from the mercury by oxidising their surface forming a thin layer of antiferromagnetic (AF) CoO. When field cooled to 77 K, a shift in the hysteresis loop by an amount H_{ex} was observed. An enhancement in the coercivity, H_c, defined as the half width of the hysteresis loop was also observed. These parameters are shown schematically in Fig. 4.1. Also shown in the figure is the training effect, ΔH_{ex}, typically observed in many, but not all, exchange bias systems. This refers to the reduction in magnitude of the loop shift upon subsequent field cycling.

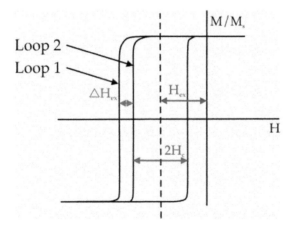

Fig. 4.1: Schematic representation of the most significant features in the hysteresis loop of an exchange bias system.

There are two components to training: an athermal one observed only in the descending branch of the hysteresis loop between loops one and two and a thermal one that affects both branches of the loop depending on the temperature of measurement. The former is indicative of changes at the ferromagnetic/antiferromagnetic (F/AF) interface (Fernandez-Outon *et al.*, 2006) while the latter is controlled by thermal activation of the bulk of the AF where both branches of the hysteresis loop move. In the figure only the athermal training is shown.

Given that CoO is antiferromagnetic below room temperature (293 K), exchange bias was attributed to an exchange coupling between the ferromagnetic Co core of the particle and the surface AF layer of CoO. In their original work, Meiklejohn and Bean also found that exchange bias was a unidirectional anisotropy as the torque curve was proportional to $\sin(\Theta)$ and not to $\sin(2\Theta)$ as in materials with uniaxial anisotropy, where Θ was the angle between the applied field and the cooling field direction. The original data of Meiklejohn and Bean is shown in Fig. 4.2.

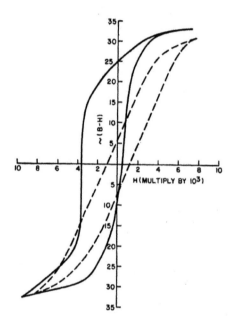

Fig. 4.2: Original data from Meiklejohn and Bean (1956) for Co/CoO nanoparticles. Reprinted figure with permission from [Meiklejohn and Bean, Phys. Rev., 102, 1413 1956] Copyright (1956) by the American Physical Society. Electronic format permissions: http://dx.doi.org/10.1103/PhysRev.102.1413

It was the development and implementation of the spin-valve sensor by IBM in 1991 (Dieny et al., 1991) that led to the number of publications in the field to increase exponentially. It is in the magnetic recording industry where the potential of exchange bias has been realised to its greatest extent where the effect is used to fix the magnetic configuration of one of the ferromagnetic layers in the giant magneto-resistance (GMR) read-head sensor while a separate (soft) free layer is left to follow the stray field from nearby bits. It is the relative orientation of the magnetic moments in these two layers that gives rise to different resistance values: high for antiparallel configuration and low for parallel (Section 5.3). The development of the tunnelling magnetoresistive sensor (TMR) in the mid 90s led to the development of magnetic random access memories (MRAM), (Chapter 8) where exchange bias can be used to establish a magnetic reference layer. More recently, exchange bias has found application in other fields such as magnetic biosensors for genetic screening (Lagae *et al.*, 2005), biosensors for DNA detection (Ferreira *et al.*, 2003) and more recently neuromorphic computing (Kurenkov *et al.*, 2020).

These days it is common to have hundreds of magnetic sensors in a car many of which make use of the exchange bias effect. For instance, many 2 dimensional angular sensors work under the same physical principle as all other GMR/TMR devices and consist of two Wheatstone bridges with four TMR/GMR sensors in each. Each bridge is sensitive to a magnetic field in orthogonal directions allowing for a full 0° to 360° range to be detected. The main difference between the GMR/TMR stacks used in angular sensors and those used in read-heads is that the former works in the saturated region while the latter operate in the linear range of the GMR/TMR characteristics.

The first AF material to be used in a GMR sensor was NiO. Given that NiO is a transition metal oxide, the nature of the AF ordering derives from the superexchange interaction where the ordering of the spins in the transition metal is controlled by the p orbitals in the oxygen atoms as discussed in Section 2.7. This material has a rocksalt structure with low magnetocrystalline anisotropy ($\sim 10^4$ ergs/cc) (Berkowitz and Greiner, 1965). Since the anisotropy controls the thermal stability of the material, read-heads consisting of a NiO layer were required to be reset every few hours. Consequently the use of NiO was soon abandoned and FeMn became the AF material of choice given its improved thermal stability due to its higher magnetocrystalline anisotropy ($\sim 10^5$ ergs/cc) (Fernandez-Outon *et al.*, 2008) arising from its fcc (111) orientation. However FeMn had its own limitations. In particular, it offered poor corrosion resistance.

Around the mid-1990s this resulted in FeMn being replaced by another Mn based fcc (111) AF material: IrMn. Given the inert properties of Ir, corrosion resistance was significantly improved and so was the thermal stability given the strong anisotropy in IrMn ($\sim 10^6 - 10^7$ ergs/cc) (Aley *et al.*, 2008). Another Mn based AF material, PtMn, has been used in a thermal MRAM device but not in a read-head mainly due to the requirement to anneal it at high temperatures to achieve the required $L1_0$ anisotropic phase (Farrow *et al.*, 1997). Although IrMn is the AF material of choice for device applications, there are some limitations with this material, the main one being its scarcity: there are only 10^{-5} atoms of Ir per million atoms of silicon, making Ir the scarcest material on Earth. Hence, there is great interest in the identification of novel AF materials that might be suitable for device applications.

Although the underlying physical processes behind the exchange bias phenomenon are complex, it is possible to construct a simple and intuitive picture of this effect. Consider that an AF/F bilayer is heated over the Néel temperature of the AF but below the Curie point of the F layer. If this is done in the presence of an applied magnetic field, the spins in the F layer will align in the direction of the field while the AF spin axes will point at random. This is shown schematically in Fig. 4.3(a). If the system is now cooled below the Néel temperature of the AF layer, still in the presence of the applied field, the exchange interaction across the AF/F interface will force the AF spin axes at the interface to align in the direction of those in the F layer still keeping AF ordering throughout the AF. This is shown in Fig. 4.3(b). If the field is now reversed, the F spins will commence to rotate. If the anisotropy in the AF layer is large enough, the AF spins will remain unchanged as shown in Fig. 4.3(c). This can be interpreted as the AF spins exerting a torque via direct exchange coupling on the F spins which will prevent them from reversing. In practice, this means that a larger field is required to reverse the F layer when grown adjacent to an AF layer. However, as the field is reversed back to its original direction, the torque from the AF spins shown as white arrows will now be acting in the same direction as the field as shown in Fig. 4.3(d). As a result the F layer reverses at a smaller or even a negative field. Overall this leads to a shifted hysteresis loop along the field axis. In this simple picture any uncompensated AF spins at the interface are not important as the alignment takes place when the AF is in a paramagnetic state.

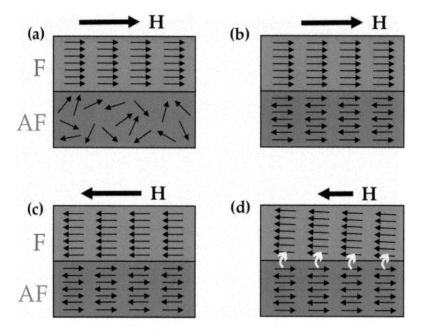

Fig. 4.3: Schematic diagram of the spin configuration of an AF/F bilayer under differ-ent field conditions: (a) after heating above the Néel temperature of the AF but below that of the F layer in a positive saturating field, (b) same field conditions after cooling below the ordering temperature of the AF, (c) after reversal of the F layer and (d) as the field is reversed back to its original direction.

Over the years, many theoretical models have been proposed all of which rely on different assumptions such as the interface spin configuration, namely whether it is uncompensated or compensated, i.e. interfaces having a net surface moment and zero surface moment, respectively. One of the main reasons why it took decades to develop a clear understanding of this phenomenon is the wide range of samples that have been studied and assumptions made by the different proposed models. For instance, it seems more than plausible to suggest that if studying samples of magnetic nanoparticles it would not be appropriate to use a model assuming perfectly flat interfaces. Similarly it is also reasonable to suggest that single crystal films grown by molecular beam epitaxy (MBE) and polycrystalline films grown by sputtering, might require different theoretical approaches. This is due to the fact that AF domains have been predicted to exist (Néel, 1948) and observed experimentally using neutron diffraction (Schull *et al.*, 1951) in single crystal AF films. However there is no evidence for the existence of AF domains in metallic polycrystalline thin films. Hence, in a similar way to ferromagnets, single domain and multi domain AFs will behave in completely different ways and will require very different mathematical treatments. Interestingly, it is single domain AF materials that show the largest exchange bias in thin film form. Hence, for applications such as read-head sensors and STT-MRAM it is metallic polycrystalline films that are used.

In the next sections we will describe some of those models. A clear distinction has been made between models developed before the turn of the millenium (Early Models) and those developed after the year 2000 (Numerical Models) when the use of large-scale computational models became readily available.

4.2 Early Models of Exchange Bias

All early attempts at explaining the exchange bias phenomenon have one thing in common: they all proposed an analytical solution. The first of these attempts was given by Meiklejohn (1962) where he assumed a perfectly uncompensated/flat AF interface. In this model, known as the rigid model because the AF moments remained rigidly aligned along their easy axis during the time of measurement, the magnitude of the loop shift was estimated to be

$$H_{ex} = \frac{J_{int}}{M_s t_F} \tag{4.1}$$

where J_{int} is the interface coupling constant, M_s the saturation magnetisation of the F layer and t_F the thickness of the F layer. The values of H_{ex} predicted by this model were typically two orders of magnitude larger than those observed experimentally. However, it is not surprising that the agreement is poor as the assumption of a perfectly flat interface is not realistic when working with ~20 nm Co particles.

Ten years after exchange bias was discovered, Néel (1967) proposed a theory that applied to a weakly uniaxial anisotropic AF layer coupled to an F layer. In his model Néel also assumed an uncompensated interface but proposed that the interfacial spin structure is subject to deformation and experiences irreversible changes during the time of measurement. Néel also considered the case of realistic rough interfaces where both compensated and uncompensated AF sublattices would be present at the interface. However, this model again failed to predict reasonable values for H_{ex}.

One of the most important models of exchange bias was that proposed by Fulcomer and Charap (1972a and 1972b). In their original work, these workers undertook both theoretical and experimental studies of exchange bias in permalloy (NiFe) films. Partial oxidation of the NiFe produced isolated islands or particles of antiferromagnetic NiO on the surface of the permalloy layer. They found that it was important to consider a distribution of particle sizes although the exact form of that distribution was not critical. Of particular significance are the predictions of the temperature and frequency dependence of the loop shift. This model has formed the basis of most other granular models based on thermal fluctuation effects.

Almost two decades later in 1987, Mauri *et al.* (1987) proposed the first domain model of exchange bias. In this model the formation of domain walls parallel to the AF/F interface above a criticial AF thickness becomes more energetically favourable. This in turn resulted in more reasonable values for H_{ex} given by eqn 4.2. However, this model cannot account for the enhanced coercivity of the F layer and the training effect. The model relies on perfectly flat interfaces and thick AF layers which are necessary to accommodate a domain wall parallel to the interface. However exchange bias has been reported in systems containing AF layers as thin as a few atomic layers. The model predicted the loop shift to be given by

$$H_{ex} = \frac{2\sqrt{A_{AF}K_{AF}}}{M_s t_F} \tag{4.2}$$

where A_{AF} is the AF exchange stiffness and K_{AF} the crystalline anisotropy. A_{AF} is a measure of the effective exchange energy within a unit cell taking into account all spin-spin interactions. It is given by

$$A_{AF} = \frac{J_{ex} n S^2}{a} \tag{4.3}$$

where J_{ex} is the exchange integral, n is the number of atoms per unit cell, S is the spin on each atom and a is the lattice parameter. The value of n equals 1 for a simple cubic array, 2 for a bcc lattice and 4 for an fcc structure. The units of A_{AF} are ergs/cm.

In the same year, Malozemoff (1987) also proposed a domain model although in his model the domain walls were perpendicular to the interface. Rough interfaces gave rise to both compensated and uncompensated regions upon field cycling. The model assumed a uniaxial single crystal F layer in contact with an AF where a domain wall within the F layer was driven by an applied in-plane field. Supposing that the interfacial energy differed for two domains, H_{ex} was determined from the balance of the applied field pressure, $2H M_s t_F$, and the effective pressure from the interfacial energy, $\Delta \sigma$, given by in eqn 4.4. This model is only applicable to single crystal AFs and does not explain the appearance of exchange bias in perfectly compensated interfaces.

$$H_{ex} = \frac{\Delta \sigma}{2 M_s t_F} \tag{4.4}$$

Ten years after Mauri and Malozemoff proposed their domain models, Koon (1997) introduced a microscopic explanation for exchange bias in systems containing compensated interfaces. The main contribution of this model was the prediction of spin-flop coupling at the interface between the F and AF layers. Koon also showed that the AF spins also exhibit canting. It was later shown by Schulthess and Butler (1998) that the model of Koon cannot result in shifted hysteresis loops but can only predict the enhanced coercivity.

In 1999 Stiles and McMichael proposed a model to describe the behaviour of poly-crystalline exchange biased systems. A critical angle was included in the model to account for irreversible transitions in the AF grains. The temperature dependence of H_{ex} was assumed to arise due to thermal instabilities in the AF grains. Based on this model, Stiles and McMichael found two contributions to the enhanced coercivity, one due to inhomogenous reversal and the other to irreversible transitions in the AF layer.

4.3   Numerical Models of Exchange Bias

Around the year 2000, microprocessor clock speeds exceeded the GHz level. This resulted in the appearance of more complicated theoretical models thanks to the development of micromagnetic and even atomistic modelling codes. Those interested in the development of our theoretical understanding of the longest running conundrum in the field of magnetism at the turn of the century, should refer to an extensive review of the state of knowledge in the year 2000 by Berkowitz and Takano (1999).

Nowak *et al.* (2002) proposed what is known as the domain state model. In this model the F layer was assumed to be coupled to a diluted AF. Dilutions were introduced in the system in the form of non-magnetic inclusions. Exchange bias then arises due to a domain state that develops during field cooling carrying an irreversible magnetisation. The dilutions in the AF favour the formation of this state since domain walls preferentially pass through non-magnetic sites reducing the energy to create a domain wall. The model did not include discretised grains and hence, cannot account for experimental features observed in polycrystalline systems. In a separate work by Scholten *et al.* (2005) the same model was applied to study the temperature dependence of H_{ex} and H_c. The main conclusion of that work was that both parameters are not related.

A few years later Saha and Victora (2006) proposed a large-scale (2.5×10^5 elements) micromagnetic model for polycrystalline systems consisting of exchange decoupled AF grains. Each F grain was coupled to an AF grain that rotated uniformly under the effect of thermal fluctuations. The Landau–Lifshitz–Gilbert equation was used to calculate the behaviour of the magnetisation in the F layer. The coupling across the F/AF interface was assumed to be due to the surface roughness of the AF grains. The authors investigated the temperature dependence of the exchange bias and the coercivity as a function of the thickness of the AF layer. An important conclusion from this model is that systems with AF symmetry higher than uniaxial, exhibit an asymmetric magnetisation reversal when compared to the uniaxial case.

All the models outlined above have been successful to some limited degree in describing one or more properties of exchange bias systems. However given the mass of data in the literature some agreement by coincidence is almost inevitable. Models based on AF domain walls can only apply to single crystal films where such walls can exist. However they have often been applied to polycrystalline films where the grain sizes are obviously too small ($\sim$10 nm) to accommodate such structures. It is then assumed that the AF grains are exchange coupled to allow the walls to exist.

Zhu and Bertram (1989) showed that the intergranular exchange coupling is proportional to the magnetisation of grains when studying the exchange coupling between neighbouring grains in ferromagnetic CoCr films with perpendicular anisotropy. For an AF film, this magnetisation might not be completely zero due to the presence of uncompensated spins at the grain boundaries, but it would be extremely small when compared to the magnetisation of an F grain where all the spins are aligned in the same direction. Hence it is reasonable to suggest that the AF grains in a polycrystalline film will be exchange decoupled. No other mechanism for AF grain coupling is known.

Another reason for the failure of the models is the wide variety of systems that have been studied: it is not plausible that a model that can account for the experimental features observed in approximately spherical core-shell nanoparticulate systems can also be applied to single crystal films containing almost perfectly flat interfaces. Furthermore, most experimental data in the literature have been taken at arbitrary temperatures without regard to possible thermal instabilities in the AF layer. Similarly the degree of order in the AF at the beginning of the measurement is generally unknown and rarely checked, which can have a significant effect on the measured magnetic properties as will be discussed in Section 4.4.1.

4.4 The York Model of Exchange Bias

4.4.1 Granular Effects

Over the last 20 years our group has undertaken a series of experiments looking to address some of the questions/issues that any proposed model of exchange bias needs to answer. These results together with their interpretation, have become what is known as The York Model of Exchange Bias and were summarised in a seminal paper in 2010 (O'Grady *et al.*, 2010). Although very simple in principle, the model has been extremely successful at predicting almost all of the experimental features observed in metallic, polycrystalline thin film exchange bias systems. Of particular interest are the AF thickness and AF lateral grain size dependence of the loop shift. This topic is very important from a technological point of view as the layers in a modern read-head sensor in a hard disk drive lie perpendicular to the plane of the disk. This means that the linear on-track density is controlled by the thickness of the layers in the stack. In the York Model the AF grains are exchange decoupled and reverse their orientation via a reversal mechanism equivalent to that observed in Stoner-Wohlfarth systems as first proposed by Fulcomer and Charap (1972a,b) and is described in Section 2.4.

One of the major successes of the model was the development of reproducible measurement protocols for the determination of the thermal stability of exchange bias systems. Any thermal instability is referred to via a blocking temperature which is directly analogous to the phenomenon of blocking in single domain particles described in detail in Section 3.2. For an AF grain it will retain its AF order but the orientation of the AF axes can be altered by an exchange field in a time dependent manner discussed in Section 3.5. Eventually the orientation may fluctuate rapidly in a similar manner to a superparamagnet.

The AF blocking temperature, T_B, is conventionally defined as the temperature at which the loop shift vanishes. At temperatures higher than T_B the hysteresis loop remains symmetrical about the field axis. This is usually determined by measuring hysteresis loops at increasing temperatures until a temperature is reached at which $H_{ex} = 0$. Fulcomer and Charap (1972a) indicated that this temperature corresponds to the blocking temperature of the AF grain with the largest anisotropy energy which in most cases will correspond to the grain with the largest volume. For magnetically decoupled polycrystalline systems it is reasonable to assume that each grain will have its own blocking temperature. Hence, the concept of a distribution of blocking temperatures arises naturally. There is one important problem with this conventional measurement procedure in that the AF is thermally unstable during the time of measurement leading to changes in its state of order during the measurement. This leads to a lack of reproducibility in the data. The values of T_B obtained in this manner will depend upon the timescale of measurement as predicted by Xi (2005) and observed experimentally by van der Heijden *et al.* (1998).

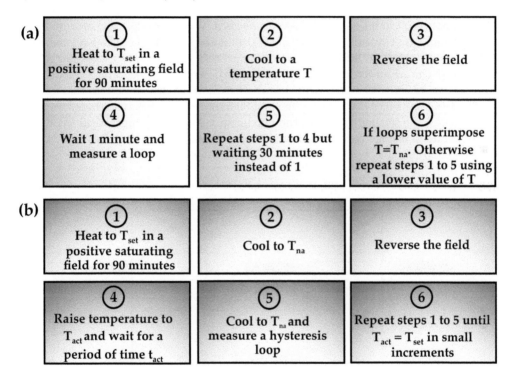

Fig. 4.4: Measurement steps in the York Protocols (a) for the determination of T_{na} and (b) controlled thermal activation of the AF layer.

The question then arises as to how to characterise the thermal stability of an exchange biased system in an accurate and reproducible manner. It should be noted that the initial alignment of the AF known as setting, cannot be undertaken above the Néel temperature for many materials such as FeMn and IrMn. For the current universally used IrMn the value of T_N is 730 K and significant diffusion will take place at this temperature. Hence setting is undertaken with the F layer saturated to generate an exchange field on the AF and the sample heated to about 500 K. The setting is then time dependent and takes one or two hours. The steps involved in these protocols which we have called the York Protocols (Fernandez-Outon *et al.*, 2004) are shown in Fig. 4.4. First, the temperature at which the AF is free of thermal activation which we call T_{na} needs to be determined. The steps for the determination of T_{na} are shown in Fig. 4.4(a). This is followed by the the controlled thermal activation of the AF layer as shown schematically in Fig. 4.4(b). Following these steps ensures that the state of order in the AF layer both prior to, and during the time of measurement that can be determined and controlled.

The AF material used in almost all applications is IrMn due to its excellent corrosion resistance and good thermal stability. The reproducible setting of the AF layer is achieved by heating the sample in the presence of a magnetic field, large enough to saturate the F layer, to increasingly higher temperatures until the highest temperature that does not result in interfacial diffusion is reached. We denote this temperature T_{set}. We have found that a setting time t_{set} of 90 minutes is sufficient to set all samples studied in our work. This is because the setting process leads to a $ln(t)$ variation in the alignment of the AF order following the processes described in Section 3.5. Figure 4.5 shows the variation of the $ln(t)$ function as a function of time. Based on this figure, 87% of the setting takes place during the first 90 minutes. Waiting an extra 90 minutes only increases the setting to 90% of the maximum value and is impractical. At these levels the rate of change is very low leading to a reproducible state of the order in the AF. Hence, 90 minutes was deemed to be a good compromise.

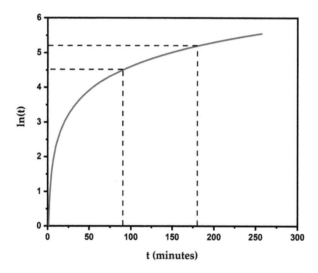

Fig. 4.5: Variation of $ln(t)$ as a function of t.

We then need to establish a temperature at which the AF orientation is free of thermal activation, which we denote T_{na}. The system is cooled to a trial T_{na} after setting, and the field reversed so the F layer is now saturated in the opposite direction to that used during the setting process. The sample is held for 1 minute in that configuration and a hysteresis loop measured. This also removes the first loop training effect. The process is repeated but increasing the time the sample is held with the F layer reversed to 30 minutes. If both hysteresis loops do not superimpose then thermal activation must have taken place during the time that the F layer was reversed and a lower value of T_{na} must be sought. This part of the protocols is shown in Fig. 4.4(a).

Once the sample is set and T_{na} is known, controlled thermal activation experiments can be performed. Rather than measuring the maximum blocking temperature, the median blocking temperature $< T_B >$ is determined. This is the temperature at which equal fractions of the volume of the AF are oriented in opposite senses. After setting and cooling down to T_{na}, the field is reversed. The sample is then heated for 30 minutes at different activation temperatures T_{act}, in appropriate steps i.e. $T_{na} \leq T_{act} \leq T_{set}$. The system is then cooled to T_{na} and the hysteresis loop measured starting from negative saturation ensuring that training effects are again removed. Since the measurements are done at T_{na} only first loop athermal training occurs. This procedure is shown in Fig. 4.4(b).

Examples of the hysteresis loops obtained following this measurement protocol are shown in Fig. 4.6(a) for a CoFe(10nm)/IrMn(5nm) exchange bias system. The resulting variation of $H_{ex}(T_{act})$ is shown in Fig. 4.6(b) with $< T_B >$ corresponding to the value of T_{act} at which $H_{ex} = 0$.

Metallic polycrystalline AF layers consist of an assembly of single domain AF grains distributed in size. The relaxation time for each of those grains is given by the Néel-Arrhenius law.

$$\tau^{-1} = f_0 \, exp - \left[\frac{K_{AF}V \, (1 - H^*/H_K^*)^2}{kT} \right] \tag{4.5}$$

where f_0 is taken to be 10^9 s^{-1}, V is the grain volume, k is Boltzmann's constant, K_{AF} is the anisotropy constant of the AF grain and T is the temperature. H^* is the exchange field from the F layer and H_K^* is a pseudo-anisotropy field similar to the anisotropy field in ferromagnets. The energy barrier can be taken to be $K_{AF}V$ as H^*/H_K^* is negligible for materials with a high value of K_{AF} (Vallejo-Fernandez *et al.*, 2008a). It is important to note that K_{AF} is temperature dependent since its origin is magnetocrystalline with a temperature dependence of the form

$$K_{AF}(T) = K_{AF}(0)(1 - T/T_N) \tag{4.6}$$

based on the dependence of K_{AF} on the magnetisation of the AF sublattice and the temperature dependence of the sublattice magnetisation (Stiles and McMichael, 1999), where T_N is the Néel temperature of the AF. At the median blocking temperature $< T_B >$, equal fractions (volumes) of the AF are oriented in opposite directions. This allows for the determination of the anisotropy constant in the AF layer. From eqn 4.5 it follows that (Vallejo-Fernandez *et al.*, 2007)

$$K_{AF} (< T_B >) = \frac{ln(\tau f_0)k}{< V >} < T_B > \tag{4.7}$$

where the grains being thermally activated at $< T_B >$ will be those having a volume $< V >$ equal to the median volume in the distribution. The measurement time τ in

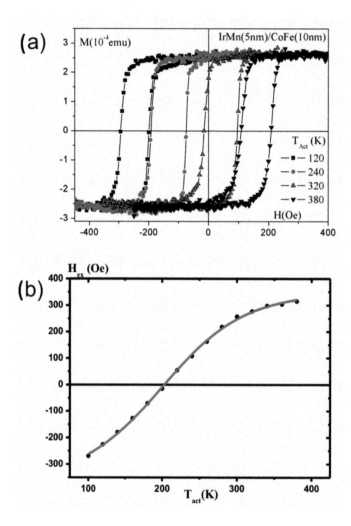

Fig. 4.6: (a) Example of hysteresis loops and (b) H_{ex} vs T_{act} and the distribution of blocking temperatures obtained for a CoFe(5nm)/IrMn(10nm) using The York Protocols (Fernandez-Outon *et al.*, 2004).

eqn 4.5 is given by the time t_{act} used for the thermal activation process in the York Protocols.

Once the value of K_{AF} is known, the effect of the thickness of the AF layer, t_{AF}, on the loop shift can be investigated. Figure 4.7 shows a schematic of the grain volume distribution in the AF layer in an exchange biased system. Since in most cases it is not possible to set at a temperature higher than the Néel temperature of the AF layer, some grains in the distribution, the bigger ones, might not be set at the chosen temperature T_{set}. This results in a maximum grain volume that will be set by the initial alignment process. This critical volume is denoted V_{set} in the schematic shown

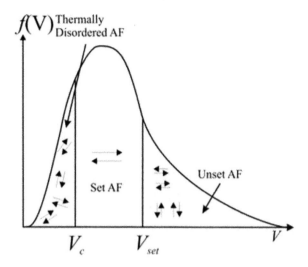

Fig. 4.7: Schematic of the grain volume distribution in the AF layer after setting at a temperature T_{set} in a magnetic field and cooling to a temperature where a fraction of the AF is thermally unstable.

in Fig. 4.7. Grains with a volume $V > V_{set}$ will not be aligned with the F layer. Similarly, there might also be a fraction of the AF grains that are thermally unstable at the temperature of measurement T_{meas} ($T_{meas} > T_{na}$). This fraction corresponds to the grains with $V < V_c$ in Fig. 4.7. Therefore, only those grains that meet the requirement $V_c < V < V_{set}$ contribute to the loop shift and, hence (O'Grady *et al.*, 2010):

$$H_{ex} \propto \int_{V_c(T_{meas})}^{V_{set}(T_{set})} f(V) \, dV \qquad (4.8)$$

However for measurements made at T_{na}, V_c should lie outside the distribution, i.e. it can be taken as zero in practice.

The proportionality sign in eqn 4.8 arises because the strength of the coupling between the F and AF layers is unknown. We denote the strength of this interfacial coupling C^*. Equation 4.8 can then be used to fit the variation of H_{ex} with a range of parameters using a constant value of C^*. This results in the scaling of the data along the ordinate but does not affect the form of the variation along the abscissa.

We have used this simple theory to explain the dependence of the loop shift on both the thickness of the AF layer and the lateral grain size. The original data for a set of samples consisting of IrMn/CoFe bilayers are reproduced in Fig. 4.8(a) and (b). These results are highly significant as their publication was the first time that a theoretical curve for exchange bias agreed with experimental data within error some 52 years after exchange bias was discovered. The error bars on the figures are from the error in integrating the particle volume distribution for use with eqn 4.8. To obtain that fit it is required to determine the volume distribution by TEM to high accuracy.

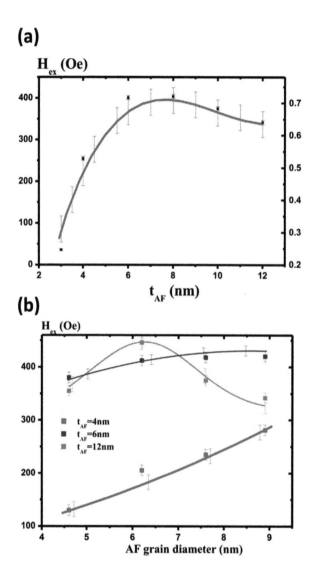

Fig. 4.8: (a) Variation of the loop shift as a function of (a) the AF thickness and (b) the lateral grain diameter for three different AF thicknesses, both measured at room temperature. (Vallejo-Fernandez *et al.*, 2008b).

For each sample 800 grains were measured using a calibrated light-box. Accordingly Fig. 4.8(a) contains almost 5000 measurements made by hand and Fig. 4.8(b) twice that number (Vallejo-Fernandez *et al.*, 2008b).

Figure 4.8(b) is highly significant because at the time it was published in 2008 there were conflicting reports in the literature that H_{ex} increased with grain size and decreased. Our model explained both possibilities.

The technological significance of the data is also clear. At the time of publication the hard drive industry believed that having smaller grains was always desirable and the thickness of the AF layer in read heads was ~12 nm. Once it was known that H_{ex} was grain volume dependent the lateral grain size was increased allowing the AF layer thickness to be reduced to ~6 nm, thereby increasing the on-track resolution significantly.

It has been acknowledged that this simple model has been used by both Seagate Inc and Western Digital Corp since 2010 in the design and manufacture of their AF materials enabling a typical 40% increase in the thermal stability of the antiferromagnetic materials used in computer hard drive read-heads (O'Grady et al., 2010). This has also resulted in a 25% improvement in the resolution of detecting a bit as in perpendicular recording the overall thickness of the read head stack determines the linear data density along the track.

4.4.2 Setting Criterion for the AF

Given that in most cases technological exchange bias structures are produced by sputtering, a wide distribution of grain sizes occurs. This in turn results in a wide distribution of energy barriers in the AF. In ferromagnetic systems, this leads to a $ln(t)$ law in the time dependence of the magnetisation. In the case of AF systems, the wide distribution of energy barriers gives rise to a $ln(t)$ variation in the setting process of the AF.

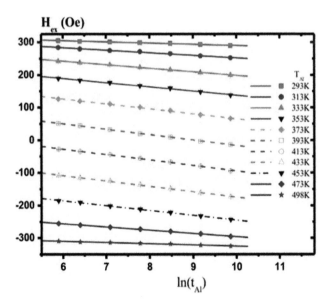

Fig. 4.9: Variation of the loop shift as a function of the setting time for different values of the setting temperature (Vallejo-Fernandez et al., 2008c).

Our group has investigated the validity of this hypothesis for a Ru(5nm)/IrMn (10nm)/CoFe(3nm)/Ta(10nm) exchange bias system (Vallejo-Fernandez *et al.*, 2008c). Prior to measurement, the state of order in the AF layer was set by heating to a temperature T_{set}=498 K in the presence of a negative saturating field, H_{set}=-1 kOe. The system was subsequently heated to different alignment temperatures T_{Al} for different periods of time t_{Al} with the F layer saturated in the positive direction (H_{set}=1 kOe). T_{Al} was varied in the range 293 K< T_{Al} < 498K since this particular system was thermally stable at room temperature. Figure 4.9 shows the variation of H_{ex} as a function of $ln(t_{Al})$ over the range of temperatures studied. H_{ex} varies linearly with $ln(t_{Al})$ for all values of T_{Al} as indicated by the quality of the linear fits. This $ln(t)$ variation of H_{ex} justifies the setting time of 90 minutes since $ln(t)$ is almost flat at this point.

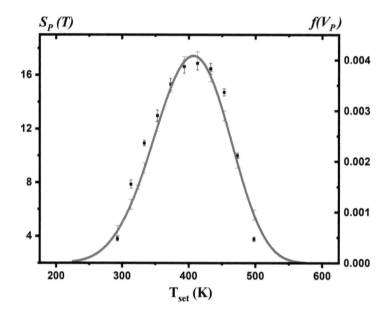

Fig. 4.10: Temperature dependence of the magnetic viscosity coefficient for a 10 nm IrMn layer. The solid line is calculated from eqns 4.9 and 4.10 and the error bars result from the error in the measurement of the grain size distribution (Vallejo-Fernandez *et al.*, 2008c).

The time dependence coefficient of the order (P) in the AF ($S_P(T)$) can now be measured from the slope of the lines in Fig. 4.9 which is shown in Fig. 4.10. The coefficient increases with increasing T_{Al} reaching a peak at the measured value of < T_B > (=413±5 K). Figure 4.10 also shows the calculated temperature variation based on eqns 4.9 and 4.10 (solid line) where $S_P(T_{Al})$ is the magnetic viscosity coefficient of the AF grains for a given value of T_{Al}, V_p is the critical volume that can be activated during the alignment process and $f(V_p)$ is the critical value of the energy barrier distribution evaluated at V_p which is given by the grain volume distribution in the sample. For the calculation a time constant $t_{Al}=120$ s was used since that was the shortest period of time spent at T_{Al} for each set of data. The calculation is again based on the measured grain volume distribution based on the work of Gaunt (1986) discussed in Section 3.5.

$$S_P(T_{Al}) \propto f[V_p(T_{Al})] \tag{4.9}$$

where $V_P(T_{Al})$ is given by

$$V_p(T_{Al}) = \frac{ln(T_{set}f_0)kT_{Al}}{K(0)(1 - T/T_N)} \tag{4.10}$$

where $S_P(T_{Al})$ is calculated from the grain volume distribution measured by TEM imaging at the various values of T_{Al}. This expression is directly analogous to eqn 3.28 which applies to ferromagnets (Gaunt, 1986). Hence the peak observed in Fig. 4.10 is a direct reflection of the grain volume distribution. Put simply the peak occurs because there are more grains of that size that are activated by the temperature.

This simple granular model has been found to account accurately for the features observed experimentally. This experiment is a critical test for The York Model of Exchange Bias since it is distinct from grain size effects. Importantly this result and the agreement with theory allows AF systems to be designed for specified setting conditions or vice versa.

4.4.3   Control of the Anisotropy of the AF

A key parameter to be considered in the design of exchange bias systems is the control of the anisotropy of the AF layer, K_{AF}. The anisotropy constant is critical in determining the thermal stability of small grains and the ability to set an exchange bias system given that the energy barrier ΔE is given by $K_{AF}V$. For the case of an IrMn system, we have undertaken a short review of the role and effect of seed layers (Cu, Ru and NiCr) on the anisotropy and hence the blocking temperature of such systems (Aley *et al.*, 2008). Grazing incidence diffraction data showed that Cu and Ru produce systems exhibiting a (111) IrMn peak indicating that some degree of the IrMn lies with [111] planes lying out of the plane of the film. However, in the case of the NiCr seed layer, the absence of a (111) peak indicated almost perfect in-plane [111] texture. Importantly, different seed layers resulted in different grain size distributions in the films which has a dramatic effect on the measured value of $< T_B >$. The values of K_{AF} were determined from data similar to that shown in Fig. 4.6 and the use of eqn 4.7. A summary of these results is shown in Table 4.1.

From these results it is clear that the degree of (111) texture controls the effective anisotropy of the IrMn layers. However the interpretation is made more complex, not only because of the effect on the average grain size in the system, but also because of the likelihood that the degree of texture of the IrMn may have a significant effect on the interfacial coupling between the IrMn and the CoFe. This implies that there is no direct correlation between the seed layer and H_{ex}. The link to $< T_B >$ depends on the product $K_{AF}V_m$ and differing seed layers produce different grain sizes when grown under the same conditions.

Table 4.1 Summary of the effect of the seed layer on the properties of IrMn/CoFe exchange biased systems.

Seed layer	D_m (nm)	σ	$<T_B>$	K_{AF} ($\times 10^7$ ergs/cm^3)
NiCr	3.9	0.42	477	3.3±0.4
Ru	6.0	0.38	386	0.94±0.06
Cu	10.7	0.37	367	0.28±0.02

IrMn has a number of unusual spin structures (Tomeno *et al.*, 1999). These are sensitive to a range of parameters including the deposition temperature but principally composition. Tsunoda *et al.* (2006) reported a variation in the exchange coupling with the composition of the alloy exhibiting an unusual "top-hat" type response. We were able to reproduce that work and extended it to investigate not only the resulting value of H_{ex} as a function of composition but also the resulting value of the anisotropy constant following the procedure described earlier (Aley *et al.*, 2011). We found that the variation of the anisotropy constant does not follow the variation of H_{ex} but rather appears to be linear over a range of compositions where the value of the exchange bias is observed to vary dramatically. This is rather peculiar, as the anisotropy increases as the alloy becomes more disordered. These results are shown in Fig. 4.11.

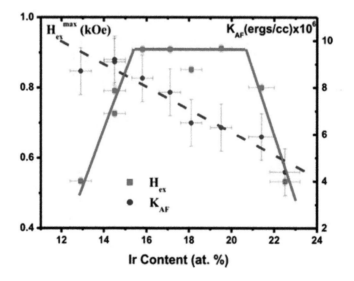

Fig. 4.11: Variation of the maximum exchange bias and anisotropy constant as a function of Ir content (Aley and O'Grady, 2011).

4.4.4 Role of Impurities and Element Size

There is significant data in the literature describing the effect of doping the AF layer on exchange bias. Notably, Fecioru-Morariu *et al.* (2007) doped the IrMn layers with Cu and found a significant increase in the exchange bias which passed through a peak at about 20% Cu. Similarly Ali *et al.* (2008) described experiments where the IrMn was produced in the form of a multilayer with Cu interlayers. Again, an initial increase in the exchange bias was observed. Similar work in our laboratory aimed to replicate these results on small grain sputtered systems suitable for applications, (Aley *et al.*, 2009). The resulting data, shown in Fig. 4.12, shows that the value of the exchange bias falls dramatically with the addition of Cu atoms reducing by 50% for only 5% impurity levels. It is difficult to reconcile these results with those for other systems. However, it is worthy of note that Fecioru-Morariu *et al.* used a system produced by MBE deposition where the grain size was of the order of 50 nm. At these sizes it is probable that the AF layer would be in a multi-domain state whereas in the small grain systems which we studied, the presence of a domain structure within the AF is inconceivable due to the lack of intergranular AF coupling.

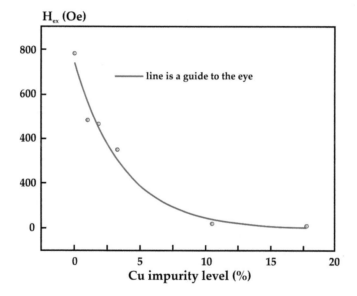

Fig. 4.12: Variation of the loop shift as a function of the Cu impurity concentration (Aley *et al.*, 2009).

This hypothesis was confirmed by further theoretical work from our group showing that the results can be understood using an AF strong domain wall pinning model (Vallejo-Fernandez *et al.*, 2011). The impurities (or defects) create pinning sites for AF domain walls in large grain or single crystal AF materials. AF domain walls are then strongly pinned at these sites increasing the proportion of single domain AF grains. This increases H_{ex} as shown in Fig. 4.13. For the calculations, a 10 nm thick IrMn layer was assumed with a median grain size of 65 nm and a standard deviation of the lognormal distribution of 0.35 so as to replicate the experimental results of Fecioru-Morariu *et al.* (2007). Increasing the impurity levels further creates more pins and reduces the effective grain size making them thermally unstable thereby reducing H_{ex}. Thus with the inclusion of pinning, The York Model accounts for the observed values of H_{ex} in the polydomain regime.

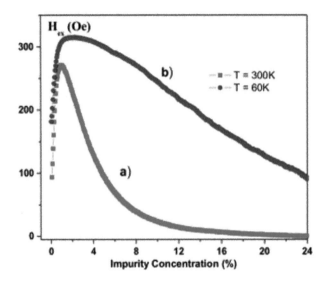

Fig. 4.13: Calculated variation of the loop shift at room temperature as a function of impurity concentration (Vallejo-Fernandez *et al.*, 2011).

Also of significant importance from a technological point of view is the dependence of the exchange bias in small elements such as read heads in hard drives and MRAM elements (see Section 7.5 and Chapter 8). Normally, the magnitude of the loop shift in small elements tends to be somewhat smaller than in the equivalent thin films. As magnetic recording technology advances, the size of a read head element is continuing to be reduced due to the requirements of scaling. This means that small elements are lithographically defined with sizes of the entire element now less than 30 nm. Similarly for applications in MRAM technology it is required to produce spin-valves or spin tunnel junctions with dimensions of the order of 20 nm in order to achieve the required storage density but particularly so if the switching of the elements is to be undertaken by spin torque switching (Section 5.5). Hence the issue of exchange bias in such small elements is highly topical and important.

Based on the York Model and the effect of the grain size on the exchange bias, one obvious possibility is that lithographic processes will reduce grain sizes at the edges of the elements leading to a different grain size distribution compared to that in a continuous film. For instance, in 125 nm elements, grains at the edges can account for up to 25% of the total area assuming a median grain size of 10 nm.

Figure 4.14(a) shows the calculated variation of the median exchange bias with element size, L, for an FeMn exchange bias system (Vallejo-Fernandez and Chapman, 2009). The inset in the figure shows experimental data by Sasaki *et al.* (2007). Figure 4.14(b) shows similar calculations assuming parameters typical of IrMn. In this case, the effect of the thickness of the AF layer was investigated while keeping the element size constant at 100 nm. As can be seen, for thick AF layers the value of H_{ex} is larger for the small elements. Increasing the thickness of the AF layer increases the grain volume, not the grain diameter. As a result, a greater fraction of the AF cannot be set as the AF thickness is increased. Upon patterning, some of those grains will be cut lowering their anisotropy energy and can therefore be set. Hence, a larger exchange bias for the nanoelements compared to the equivalent thin film is predicted for thicknesses of the AF >10 nm. These results are in excellent agreement with results from the group of Bernard Dieny (Baltz *et al.*, 2005) where the same trend was observed in IrMn/NiFe continuous thin films and sub <100 nm round elements.

4.4.5 Interfacial Effects

There has long been a controversy concerning the role of interfaces and the bulk of an AF in determining the value of H_{ex}. Generally interfaces are very difficult to study directly and remote techniques such as XMCD have to be used. However we have been able to elucidate some information concerning the behaviour of interfaces in exchange bias systems because of our ability to understand and predict the behaviour of the bulk of the AF grains. In an initial study Dutson *et al.* (2007) investigated a trilayer system consisting of two F layers of different thickness separated by a thin AF layer. This system gives rise to two hysteresis loops exchange biased to a different degree with a plateau region in-between. Hence, the York Protocols can now be applied with either one or both of the F layers oriented in the opposite direction to that in which the AF was originally set. Figures 4.15(a) and 4.15(b) shows these results.

When the system is thermally activated with both F layers reversed, the fraction of the bulk of the AF that progressively reorients increases with increasing activation temperature as expected. As a consequence, the hysteresis loop for both F layers shift to the right along the field axis in a similar manner to that in single-layer systems (Fig. 4.15(a)). However, when the system is thermally activated with just the thicker F layer reversed (Fig. 4.15(b)), thermal activation at moderate temperatures (<150°C) shifts only the loop for the thicker F layer while the loop shift for the thinner F layer remains unaltered. This can only be due to the interface between the AF and the thicker F layer being affected during the thermal activation process. If the bulk of the AF were modified, then the hysteresis loop for the thinner F layer would also move.

This surprising result is a clear confirmation of a quasi-independent spin structure at the interface. The coincidence in the shift of the thinner layer is almost exact and this could not occur if the bulk of the AF were at all affected by thermal activation with the thicker loop reversed. Given that the thickness of the AF layer was only 5 nm, it is inconceivable that a complex spin rotation could exist within the AF grains. Given the size of an Ir atom, the AF layer is of the order of 10 atoms thick which

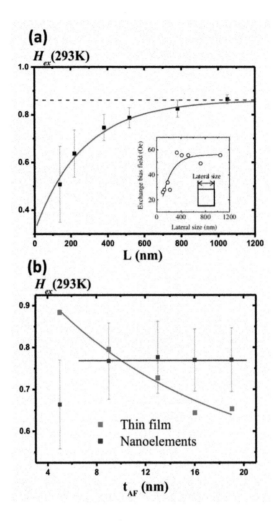

Fig. 4.14: (a) Variation of the median exchange bias field with element size L for FeMn exchange bias systems (inset reprinted from Journal of Magnetism and Magnetic Materials, 310, Sasaki *et al.*, Size effects on exchange bias in polycrystalline Ni–Fe/Fe–Mn square dots, 2677-2679, Copyright (2007), with permission from Elsevier) and (b) variation of the exchange bias as a function of t_{AF} for nanoelements (blue symbols) and thin films (red symbols) (Vallejo-Fernandez and Chapman, 2009).

would imply that the interfacial spin structure cannot extend through more than one or two atomic layers.

Further evidence of the different nature of the bulk and AF interface is given by the field setting dependence of the exchange bias. It is commonly accepted that it is the exchange field from the F layer that aligns the AF grains. This would suggest that setting a sample in a field of 500 Oe or 10 kOe should make no difference as the F layer saturates in a few hundred Oe. However, that is not the case. Figure 4.16(a)

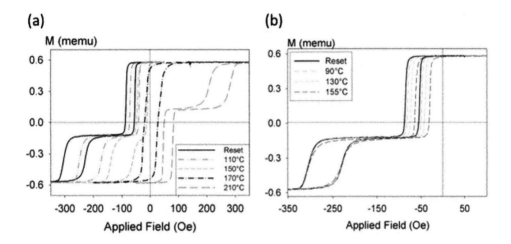

Fig. 4.15: Thermal activation of a trilayer with (a) both F layers and (b) only one F layer reversed during the thermal activation process at a variety of temperatures (Dutson *et al.*, 2007).

shows the variation of the exchange bias in the same trilayer system used earlier. An increase in the exchange bias of approximately 20% is observed for both F layers when the setting field increases from 1 to 20 kOe. The exact shape of the variation is not identical suggesting that it is the interfaces that are being changed rather than the bulk and that the spin structure in the two interfaces are not identical. Since both F layers are fully saturated by 350 Oe, changes in H_{ex} for setting fields beyond this point have to be exclusively due to changes in the interface spin order (O'Grady *et al.*, 2010).

This suggestion is confirmed by the measurement of the distribution of AF blocking temperatures for the trilayer system using different values of the setting field using the York Protocols. Figure 4.16(b) shows that the distribution of blocking temperatures is identical in all cases indicating that the bulk of the AF has not been altered by using different values of H_{set}. This confirms that the setting field effect is entirely due to interfacial spins and the effect of their order on the coupling across the F/AF interface. Again this result shows the quasi-independence of the interface spin-structure. It is also clear that there have to be spin clusters at the interfaces as shown schematically in Fig. 4.17. A single spin would be little affected by an applied field of 20 kOe particularly at the elevated temperature at which the AF is set. From these results it appears that the interface spins behave as a quasi-independent layer of spin clusters sitting between the F and AF layers.

We have undertaken measurements at low temperatures below the temperature T_{na} at which we have established that no thermal activation of the bulk of the AF grains occurred (Fernandez-Outon *et al.*, 2004). We observed that when measurements of H_{ex}

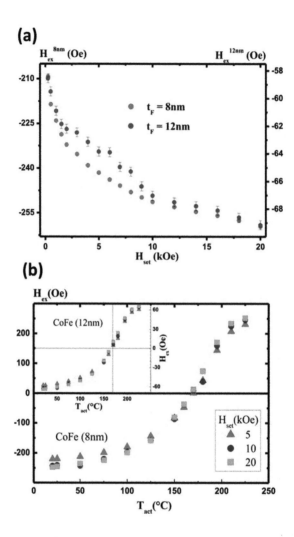

Fig. 4.16: (a) Variation of H_{ex} with H_{set} for the two CoFe layers in the trilayer system and (b) blocking temperature curve for different values of H_{set}. Reprinted from Journal of Magnetism and Magnetic Materials, 322, O'Grady *et al.*, A new paradigm for exchange bias in polycrystalline thin films, 883, Copyright (2010), with permission from Elsevier.

were made below T_{na} a progressive increase in the value of H_{ex} occurred (Fernandez-Outon *et al.*, 2008). We observed this effect in both IrMn samples but particularly in FeMn systems where the increase in the anticipated value of exchange bias commenced at temperatures of around 77 K (Fernandez-Outon *et al.*, 2008). The nature of this anomaly was confirmed by the fact that the resulting value of the exchange bias at temperatures near to 4.2 K could be as much as 50% greater than that observed when

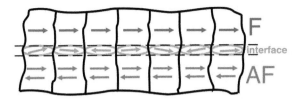

Fig. 4.17: Schematic diagram of the spin clusters present at the AF/F interface.

the bulk of the AF is fully saturated at higher temperatures. These results are shown in Fig. 4.18. We interpret this result as being due to spontaneous ordering of the interfacial spin clusters that can only occur when the sample is field cooled as in a classical blocking temperature experiment for a ferromagnet. Since the increase in the exchange bias cannot be due to a bulk effect it must therefore be due to interface spin freezing. However, given that the increase in ordering of the interfaces which transmit the exchange interaction from the AF to the F occurs at such high temperatures, this is indicative of cooperative behaviour among the interface spin clusters.

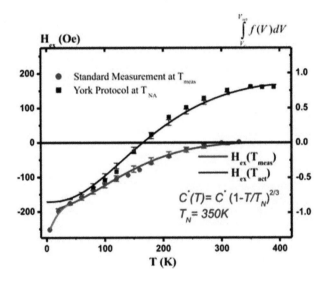

Fig. 4.18: Exchange bias field measured following the York Protocols and measured via the standard procedure for FeMn(10nm)/NiFe (Fernandez-Outon *et al.*, 2008).

Given the temperature and field dependence of the interfacial setting, it is possible to speculate on the nature of the interface spin structure. The first thing to note is that since H_{ex} increases with increasing setting field, it is reasonable to assume that the order of the interfacial spins is ferromagnetic in nature. This is consistent with the spin ordering observed at low temperatures (Fig. 4.18). Experimental measurements from our group indicate that these spins exist in clusters containing 10-50 spins (O'Grady *et al.*, 2010).

4.4.6 Interfacial Doping

Another interesting feature is that introducing an ultra-thin layer of Mn at the interface between the F and AF layers can result in an increase in the value of the loop shift when compared to a system with no additional Mn (Tsunoda *et al.*, 2007). The question then arises as to whether this effect is interfacial or in the bulk of the grains. It is known that Mn diffuses into CoFe during the high temperature setting process leaving the interface Mn deficient (Kim *et al.*, 2003). It is then plausible to assume that the increase in H_{ex} is due to the replenishment of Mn at the F/AF interface. We have undertaken a study in our laboratory where we investigated the effect of adding a thin (0 to 5 atomic layers) Mn layer at the IrMn/CoFe interface (Carpenter *et al.*, 2012). A 10 nm IrMn layer was used to ensure the bulk of the AF was thermally stable at room temperature. Figure 4.19 shows the variation of the loop shift as a function of Mn doping. The addition of 1 or 2 atomic layers of Mn results in an increase in the loop shift of ~100 Oe. Thicker Mn layers result in a large decrease in H_{ex} of up to 50%.

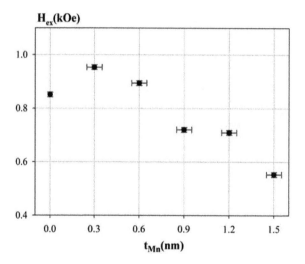

Fig. 4.19: Variation in H_{ex} due to the addition of 0 to 5 atomic layers of Mn at the IrMn/CoFe interface (Carpenter *et al.*, 2012).

There are several possible explanations for the dependence shown in Fig. 4.19. Based on work from our laboratory (Gompertz *et al.*, 2022) we believe the formation of CoMn alloys at the F/AF interface is the most plausible explanation. Men'shikov (1985) studied the magnetic behaviour of $Co_{1-x}Mn_x$ alloys when $0 < x < 0.5$. The magnetic behaviour was found to be strongly dependent on both the Mn content and temperature. It is reasonable to assume that the the composition of the clusters will vary, i.e. the Co content will be higher nearer the interface with the CoFe layer while the Mn content will be higher nearer the IrMn interface. CoMn transitions from F to

AF as the concentration of Mn increases from 0 to 50%. As a result both F and AF clusters will exist distributed in a matrix of weakly coupled particles with spin glass like properties. By changing the interfacial Mn layer thickness, the nature of the spin clusters will change. The fact that the peak in H_{ex} as a function of the interfacial Mn is less than the atomic diameter of Mn, suggests that the interfacial composition driven by interdiffusion of Co and Mn plays a significant role.

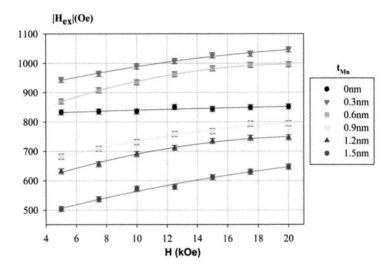

Fig. 4.20: Variation of H_{ex} with setting field for various thicknesses of the Mn insertion layer (Carpenter *et al.*, 2012).

In order to confirm that this is an interfacial effect, the field setting dependence of the exchange bias was investigated as a function of the setting field for these samples. As shown in Fig. 4.20, H_{ex} is almost constant for the sample with no added Mn. However, a field as high as 20 kOe is not large enough to fully set the interfacial clusters in the samples with Mn insertion layers as can be seen from the lack of saturation in the data in Fig. 4.20. It is reasonable to assume that the number of spins in a given cluster increases with the addition of Mn at the interface. The degree of order within these clusters controls the coupling between the F and AF layers.

One final feature is the asymmetry in the saturation magnetisation in polycrystalline exchange bias systems. These asymmetries has been predicted theoretically (Nowak *et al.*, 2002) but had not been confirmed experimentally until recently (Gompertz *et al.*, 2022). By asymmetry we refer to a difference in the saturation magnetisation between positive and negative saturation. Figure 4.21 shows the temperature dependence of this asymmetry for samples with composition shown in the inset of the figure. The figure shows data for the first (i) and second (ii) hysteresis loop measured immediately after setting. For $T > 100$ K, the asymmetry is very modest. We believe this is due to the interface spins not being ferromagnetically ordered and hence not responding to the relatively small applied fields used to measure the hysteresis loops. As the temperature is reduced, the level of asymmetry increases dramatically peaking at ~6 K as shown in Fig. 4.21 where ΔM_s is the difference in the saturation moment in positive and negative fields. This behaviour is consistent with the idea of the interface spins existing in clusters similar to a spin glass as hypothesized earlier.

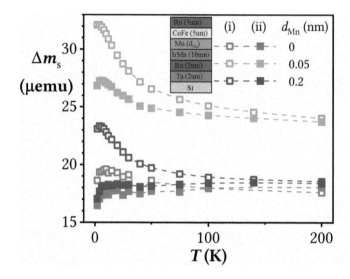

Fig. 4.21: Variation of H_{ex} with setting field for various thicknesses of the Mn insertion layer (Gompertz *et al.*, 2022).

Taking all these results into consideration, eqn 4.11 provides a simple expression that can now be used to account for most, if not all, experimental features controlled by the bulk of the AF layer. Based on the previous experimental observations on the role of the interfacial spins, we can rewrite that equation as

$$H_{ex} = H_{ex}^{i} C^{*}(H_{set}, T) \int_{V_c(T_{meas})}^{V_{set}(T_{set})} f(V)\, dV \qquad (4.11)$$

H_{ex}^{i} is an intrinsic exchange bias that would occur for a perfectly stiff interface, i.e. $C^{*} = 1$, coupling a fully set and thermally stable AF to the F layer. From the setting field experiment as well as the temperature dependence of the interfacial spin order, it is clear that C^{*}, which represents an interface stiffness, will depend both on the field used to set the system as well as the temperature of measurement. Overall, the York Model of Exchange Bias provides a physically realistic model of exchange bias in metallic polycrystalline systems based on an independent AF grain volume model.

5
Magneto-Resistance

It is an interesting fact that the first observation of magneto-resistance in a magnetic material was made by William Thomson (latterly Lord Kelvin) and reported in 1857. This observation predates the formal discovery of the electron by J. J. Thomson by about 40 years. In bulk materials this effect is known as anisotropic magneto-resistance and its origins are relatively easy to discern.

5.1   Anisotropic Magneto-Resistance (AMR)

In a completely demagnetised material electrons will be scattered evenly irrespective of the direction of the spin on the electrons forming the current density. However if the material is now saturated then all the electrons will to some degree experience a Hall effect which will cause them to circle around the direction of the magnetisation. That orbit will result in significantly longer electron paths resulting in an increased probability for scattering thereby increasing the effective resistance of the material. However those electrons whose direction of motion do not form a strong cross product with the applied field will not be scattered to the same degree. This arises due to the force on an electron in a magnetic field $\mathbf{B}$ which is due to the applied field and the magnetisation of the material

$$\mathbf{F} = e\,(\mathbf{v} \times \mathbf{B}) \tag{5.1}$$

where $\mathbf{v}$ is the velocity of the electron. This force results in a voltage being generated perpendicular to the direction of the current flow given by

$$V_H = \frac{R_H i B}{t} \tag{5.2}$$

where V_H is the Hall voltage, R_H is known as the Hall coefficient for the material under study and t is the thickness of the conductor. In general these effects are small and the magneto-resistance is given by

$$\left(\frac{\Delta R}{R}\right)_s (\theta) = \frac{3}{2} \left(\frac{\Delta R}{R}\right)_i [1 - cos^2\theta] \tag{5.3}$$

where θ is the angle between the current density and the direction of the magnetisation in the material, $(\Delta R/R)_s$ is the saturation value of $(\Delta R/R)$ at the angle θ and $(\Delta R/R)_i$ is the change observed from the ideal demagnetised state to saturation in the field direction.

In general the value of anisotropic magneto-resistance is very small. For example in Ni it is of the order of 2.5%, for Fe 0.8% and for Co in an hcp phase it is 3% along the c-axis. It should be noted that these values for magneto-resistance are those for single crystal materials. However for the resistivity of amorphous materials there is no significant effect due to the lack of long-range crystallographic order. The increased grain boundary scattering, if very small grains are present, swamps any magneto-resistive effect that may be present.

Despite these very small values of magneto-resistance it is worthy to note that early magnetic recording systems and in particular those used in early hard disk drives, did use magneto-resistive thin film read heads due to the fact that they could be defined by the relatively crude lithographic processes available at that time. None-the-less they gave a higher resolution read signal than the previously used horseshoe configurations which were also used for writing data to a disk. However with the advent of the discovery of giant magneto-resistance (GMR) and subsequently tunnelling magneto-resistance (TMR) there is almost no use at all for materials exhibiting AMR.

To understand GMR and TMR which are fundamentally quantum mechanical effects, it is first a requirement to understand the basics of the band theory of ferromagnetism.

5.2   The Band Theory of Ferromagnetism

The first significant attempts to explain the origins of ferromagnetism other than the explanation of the behaviour of ferromagnets via the concept of domains originally due to Pierre Weiss, was the molecular field theory originally proposed but not significantly developed by Alfred Ewing as noted in Chapter 2. However the significant development of the molecular field theory was also undertaken by Pierre Weiss (1906) partly in collaboration with Paul Langevin who is best known for the classical theory of paramagnetism.

The basis of this theory is that the magnetic moments on atoms in the ferromagnetic metals align with each other producing ferromagnetism via a molecular field (H_m) proportional to the magnetisation of the material. Due to the nature of this hypothesis the field would be due to some form of dipole-dipole interaction. The molecular field is then given by

$$\mathbf{H}_m = \gamma \mathbf{M} \qquad\qquad (5.4)$$

where γ is called the molecular field constant.

This theory was developed in the early part of the 20th century when of course little was known or suspected about quantum mechanics. The theory did achieve some success in that it was able to explain the form of the magnetisation versus temperature relationship leading to a Curie point. In fact a common line for this variation for the three transition metal ferromagnets was developed when the angular momentum giving rise to the magnetic moment was set to $J = 1/2$. The use of the total angular momentum was also able to explain the non-integer values for the observed number of Bohr magnetons on each atom but the agreement between calculated and measured values was generally very poor. However the greatest failure of this model was the resulting values of the molecular field itself which were generally required to be of the order of mega-Oersteds. Clearly this is non-physical even at the interatomic separation typically found in crystals.

It was only with the development of quantum mechanics and the understanding of the concept of energy levels that a true understanding of how spins within ferromagnets can order was achieved. In addition to the concept of energy levels it was also required that a knowledge and understanding of the exchange interaction (Section 2.2.1) was established.

We must now consider the origin of the bands which occur in all solid materials and give rise to the phenomenon of ferromagnetism. We are all familiar with the concept of energy levels in an atom which can be used to define the atomic structure of all atoms. However the sharp spectral lines that each atom generates due to transitions between energy levels, are transitions that are observed when a solid material is excited within a spectral lamp by striking an electric arc which results in the production of a monatomic gas of the material. For example the well known spectral series of hydrogen is that of an hydrogen atom whereas without the production of special spectral lamps, normal hydrogen gas would give the spectrum of the hydrogen molecule which is different. Hence the well known atomic spectra refer only to a monatomic gas of an element.

When atoms of a material which is normally solid at room temperature come close together for example to form a crystal structure, restrictions on the energy levels which the electrons may occupy are brought in by the Pauli Exclusion Principle. The normal definition of the Pauli Exclusion Principle states that no two electrons within an atom can have the same four quantum numbers. An alternative statement of the Pauli Exclusion Principle would be that no two electrons in the same region of space can have the same four quantum numbers when their separation is such that the electron probability clouds begin to overlap. That is always the case in crystalline structures.

Considering the transition metal elements where the three ferromagnetic metals lie, the maximum occupied shell is the 3d level but as is well known in the theory of atomic structure, the energy of the 3d level overlaps closely with the 4s level and in a number of elements the ground state sees the 4s level begin to be occupied before the 3d level is complete. That is the case for the three ferromagnetic metals and the occupancy of the 3d and 4s levels and the total number of electrons contained within them are shown in Table 5.1.

Table 5.1 Electronic structure in the 3d transition metals.

Element	Number of 3d electrons	Number of 4s electrons	Total 3d+4s electrons
Mn	5	2	7
Fe	6	2	8
Co	7	2	9
Ni	8	2	10
Cu	10	1	11
Zn	10	2	12

As a crystal of one of these metals is formed, the separation of the atoms becomes such that the electron clouds overlap whereupon the Pauli Exclusion Principle says that each individual level can hold only two electrons with oppositely aligned spins. Hence the only way that the crystal can form is for the level which would normally be discrete in the monatomic gas, to break up into a series of levels that form a band. Depending on the separation there can be tens or even hundreds of levels each of which can be occupied by two oppositely aligned electrons. This is shown schematically in Fig. 5.1. For levels closer to the nucleus the degree of band formation is significantly less. For example for the 2p this would not significantly change any aspect of the magnetism because the full occupancy of the 2p level would result in electrons with opposite spin lying within the discrete levels and not contributing to a net moment.

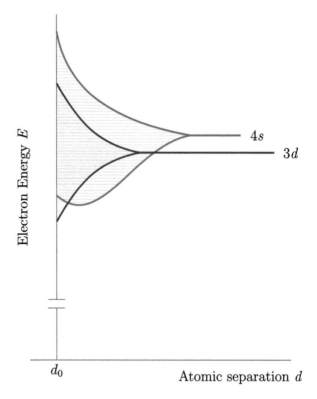

Fig. 5.1: Schematic of band splitting where d_0 is the interatomic spacing in the crystal.

Within the bands that form the 3d and 4s levels, the critical parameter to be considered is the density of available electron states into which an electron can move. The calculation of a density of states curve which is a plot of $N(E)dE$, i.e. the number of states with energy between E and $E + dE$, against energy, is highly complex but suffice to say that the density of states as a function of energy is never a smooth function and tends to be an erratic curve with a number of peaks and troughs often as many as six peaks separated by troughs. Fig. 5.2 shows a schematic diagram of such a curve taken from the text by Cullity (1972).

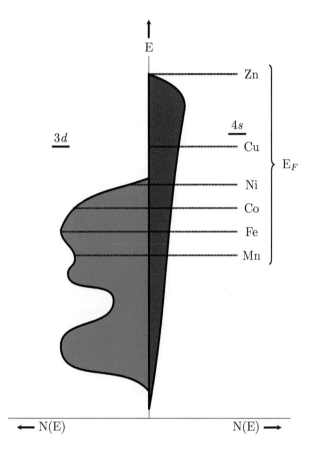

Fig. 5.2: Schematic diagram of the density of states in the 3d and 4s bands (reproduced from Cullity, © 1972 John Wiley & Sons, Inc.)

In Fig. 5.2 the density states are shown for both the 3d and 4s bands and also shown is the degree to which they are filled (not to scale) for both bands. Because of the overlap of the two bands each must be filled to the same level and that level is designated by the dashed lines which represent the Fermi level for the elements shown in Table 5.1. Of course the density of states curve for each element will to some degree be different but the basis of the argument about how this leads to ferromagnetism in Fe, Co and Ni does not depend upon the shape of the density of states curve whose variation is relatively minor. This concept is known as the rigid band approximation. As can be seen from Table 5.1 the true ferromagnetic elements all have two electrons in the 4s band and hence many of the available levels in the band itself will be unoccupied but each occupied level containing two electrons will not contribute to the net magnetic moment on the atom as their spins will be anti-parallel. Hence it is only the partially filled levels in the 3d band that contribute to the ferromagnetic order.

Given that the energy levels are separated only by a very small amount, the effect of the exchange interaction results in the fact that it is slightly energetically more favourable to have electrons in the plethora of 3d levels with their spins in a parallel orientation. The degree of favourability for this configuration is determined by the density of the levels generally called the density of states, $N(E)dE$, that determines the degree of spin alignment that occurs in the 3d band. It is then this spin imbalance that results in ferromagnetic order. It is for the same reason that the relatively low density of states in the 4s band that results in the antiparallel alignment of those electrons. For Cu and Zn the 3d band is not split and hence there is no spin ordering. The positive value of the exchange integral J_{ex} (see Fig. 2.5) coupled to the high density of states in the d-band of Fe, Co and Ni and consequently, their small separation, gives an overall lowering of the total energy of the system.

It is now necessary to look at the overall occupancy of the 3d and 4s bands. If we allow the total number of 3d and 4s electrons per atom to be represented by n and the number of 4s electrons available be represented by x then the quantity $(n - x)$ will be the number of 3d electrons available from that atom. Given the maximum number of 3d electrons is 10, the magnetic moment per atom is therefore

$$\mu_H = [5 - (n - x - 5)]\,\mu_B = [10 - (n - x)]\,\mu_B \tag{5.5}$$

where μ_B is the Bohr magneton. Taking Ni as an example then the total number of 3d and 4s electrons is 10 and we know from experiment that the value of the magnetic moment μ_{Ni} is 0.60 μ_B. Substituting these values into eqn 5.5 we find that x, the number of 4s electrons, is equal to 0.6. Given the relative lack of variation in the density of states in the 4s level shown schematically in Fig. 5.2, it is reasonable to assume that the other ferromagnetic elements which neighbour Ni in the periodic table would have a similar occupancy in the 4s band. Hence we can then calculate that the magnetic moment per atom for Fe and Co as 2.22 and 1.72, respectively. This agrees to a reasonable extent with the experimentally measured values of 2.60 and 1.60. Hence it is the spin split bands and the effect of the exchange interaction between the electrons that gives rise to the ferromagnetic order observed. This concept also explains the non-integer values for the magnetic moments on these three common ferromagnetic elements. The model has also been found to give reasonable values for a wide range of alloys of the three ferromagnetic elements.

Until relatively recently it was extremely difficult to calculate the band structure of transition metal elements. Following the development of complex computer models of band structure using techniques such as Density Functional Theory (DFT), it is now possible to produce relatively accurate density of states curves and the agreement between theory and experiment is significantly improved. Examples of such calculated band structures and a detailed discussion of the band theory of ferromagnetism can be found in a number of texts but most recently in that due to R. C. O'Handley (2000).

5.3   Giant Magneto-Resistance (GMR)

With the development of quantum mechanics and in particular the concept of electron spins in the early part of the 20th century, it was realised as long ago as the 1930s that there could be significant scattering of conduction electrons due to the presence of electron spin when they pass through a magnetised region of a ferromagnetic material (e.g. see Englert, 1932). This topic was also studied in some detail by Sir Neville Mott in Cambridge (1936) who published an almost defining work on the subject in 1936 in which he described the existence of two resistive channels in a ferromagnet which depend on the orientation of the electron spin. This concept of two different resistive channels led to significant further work on spin dependent scattering in ferromagnets most notably by Fert and Campbell (1968 and 1976).

It should be noted that the three common metallic ferromagnets are all transition metals and have higher resistivity than noble metals such as gold. This level of resistivity is due to the availability of unoccupied states in the partly filled d bands into which the electrons can be scattered which reduces the effective mean free path. As discussed in Section 5.2 there is a different density of states at the Fermi surface for spin up and spin down electrons and therefore a difference in the scattering of those electrons. In general the scattering is larger for the minority electrons in the band.

GMR was discovered simultaneously and independently by Peter Grünberg of the Forschungszentrum Julich (1989) and Albert Fert at the Université Paris-Saclay (1988) although the paper by Grünberg and co-workers did not appear until 1989 awaiting the issue of a patent on the effect. Both groups worked on thin film multilayers which in the case of Grünberg were a very simple Fe/Cr/Fe tri-layer whereas Fert and co-workers used a multilayer stack of the same elements but with between 30 and 60 repeats (Baibich *et al.*, 1988). Hence the magnitude of the effect observed by the group of Fert was significantly larger than that observed by Grünberg. Both groups studied the effect at low temperature.

The first observation made by Grünberg was that the two ferromagnetic layers in his structure automatically aligned with the magnetisation in opposite directions which led to the discovery and development of what was subsequently called a synthetic antiferromagnet (SAF). Hence he observed two distinct hysteresis loops separated by a field of the order of 600 Oe (Binasch *et al.*, 1989). Based on the discovery of GMR Peter Grünberg and Albert Fert were awarded the Nobel Prize for Physics in 2007.

Based on the two-channel model originally due to Mott, this became an obvious system in which to look at electron spin scattering. The macroscopic understanding of GMR can now be envisioned very simply through a resistor network model as shown in Fig. 5.3. Figure 5.3(a) shows the low resistance state when the two layers are oriented parallel to each other and Fig 5.3(b) the high resistance state when they are aligned in opposite directions, i.e. Fig. 5.3(a) is when the two layers are at saturation due to the application of the field and also the zero field configuration when the layers align antiparallel. Of course in both configurations there exists the two channels referring to electrons with spin up and those with spin down. Hence a parallel array of two resistors in series gives the total resistance for the system as a whole. In some cases the value of $\Delta R/R$ uses the value of R parallel but given that the antiparallel state is the natural state due to the synthetic antiferromagnetism, it is more correct to show the change in resistance being that from the normal state of the multilayer.

Pictorially one can think of the GMR effect as electrons from the majority channel have intrinsically a lower level of scattering and become polarised in the first layer they encounter. If they then encounter the second layer which is not parallel to the first they will suffer significant scattering, whereas when the two layers are aligned

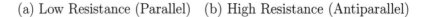

(a) Low Resistance (Parallel)   (b) High Resistance (Antiparallel)

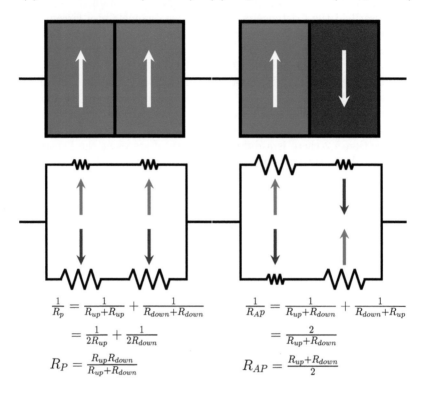

$$\frac{1}{R_p} = \frac{1}{R_{up}+R_{up}} + \frac{1}{R_{down}+R_{down}} \qquad \frac{1}{R_{AP}} = \frac{1}{R_{up}+R_{down}} + \frac{1}{R_{down}+R_{up}}$$

$$= \frac{1}{2R_{up}} + \frac{1}{2R_{down}} \qquad = \frac{2}{R_{up}+R_{down}}$$

$$R_P = \frac{R_{up}R_{down}}{R_{up}+R_{down}} \qquad R_{AP} = \frac{R_{up}+R_{down}}{2}$$

$$\therefore \Delta R = \frac{R_{up}+R_{down}}{2} - \frac{R_{up}R_{down}}{R_{up}+R_{down}}$$

$$\therefore \frac{\Delta R}{R_{AP}} = 1 - \frac{R_{up}R_{down}}{R_{up}+R_{down}}$$

Fig. 5.3: (a) Low resistance and (b) high resistance state for a resistor network model. The arrows denote spin up and down channels.

then little scattering occurs and the resistance therefore falls.

There was also around the time of the discovery of GMR, an ongoing debate as to whether the scattering took place within the bulk of the multilayer or at the interface with the non-magnetic layer. Stuart Parkin of IBM undertook an elegant experiment in which he included very thin layers of ferromagnetic metals at the interface between the ferromagnet and the non-magnetic sandwiches (Parkin, 1991). Spin dependent scattering arises from the band structure, for example if the electron potentials match in one spin direction and not the other. The results of his work show that GMR was in fact an interfacial scattering effect and he undertook significant work on the CoCu system because there is a good lattice match resulting in the low dislocation density at the interface and consequently low spin independent scattering (Parkin *et al.*, 1991).

He also observed very high values of GMR in this system at room temperature with values approaching 40% being observed when the oscillatory coupling that gives rise to the synthetic antiferromagnetic behaviour was at a maximum where the thickness of the Cu layer was below 1 nm. The oscillatory coupling is shown in Fig. 5.4.

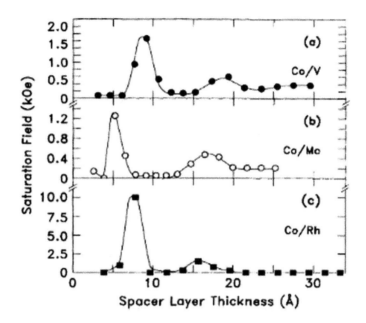

Fig. 5.4: The oscillations in the RKKY coupling strength through various spacer materials, showing the peak for Vanadium at 0.9 nm. Reprinted figure with permission from [Parkin, Physical Review Letters, 67 3598, 1991]. Copyright (1991) by the American Physical Society. Electronic format permissions: 10.1103/PhysRevLett.67.3598

It should be noted here that the majority of these early experiments were undertaken on single crystal systems. However single crystal systems are not suitable for applications due to the cost of deposition and a lack of uniformity in the magnetic behaviour due to domain wall pinning effects as noted in Section 2.3. Modern thin film deposition techniques including sputtering, have now advanced to the point where highly epitaxial growth between layers of different materials can be achieved and all modern devices in which GMR sensors are grown are generally produced by some form of sputtering or other ion beam deposition techniques.

Following the discovery of GMR an enormous effort both experimental, theoretical, academic and industrial, was immediately brought to bear to achieve the potential of this new sensing principle whose sensitivity was at least one order of magnitude greater than the thin film read heads used at that time in tape and disk storage systems. In a relatively short period of time workers at IBM developed what became known as the spin-valve sensor (Dieny *et al*, 1991).

A spin-valve is basically a simple tri-layer GMR system but with the addition of a layer of an antiferromagnetic material beneath the stack. As discussed in Chapter 4 when such a system is field annealed this resulted in a fixed hysteresis loop so that one layer of the GMR stack had its orientation fixed. Furthermore to optimise the GMR response to the very small field resulting from a bit of stored information on a thin film disk or particulate tape, biasing permanent magnet materials were incorporated either side of the stack so that the orientation of the free layer was as close to 90° as possible. With the free layer oriented in this way a spin-valve sensor is capable of detecting very tiny fields.

The reason for biasing the free layer is that if it were antiparallel to the pinned layer then the field that is being sensed has to overcome the total AF coupling in the SAF structure discovered by Grünberg (Binasch *et al.*, 1989). If the free layer is biased by 90° (if possible) very small fields will generate a signal even from small deflections. The GMR signal $\Delta R/R$ is given by

$$\frac{\Delta R}{R} = \mathbf{M}_F \cdot \mathbf{M}_P = M_F M_P cos\theta \qquad (5.6)$$

where M_F and M_P are the magnetisations of the free and pinned layers respectively and θ is the angle between them. The response as a function of θ is shown in Fig. 5.5 which is a simple cosine curve.

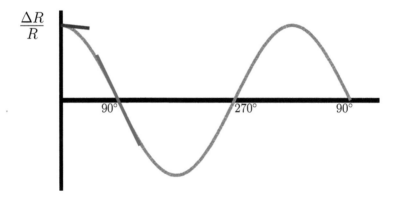

Fig. 5.5: GMR response as a function of the angle θ between the free and pinned layers.

As can be seen from the figure the $cos\theta$ curve close to 90° gives a quasi-linear varia-tion over quite a large range of θ and at a far greater level of sensitivity $(d(\Delta R/R)/d\theta)$ than is the case closer to $\theta = 0$.

The biasing is achieved by growing small areas of a permanent magnet material, typically FePt, either side of the device when the current through the device is in the plane of the layers. This generates a field strong enough to deflect the moment of the free layer but not strong enough to deflect that of the pinned layer which is kept aligned in the direction of the current flow by the exchange bias induced by the AF layer.

Equation 5.6 also indicates the requirement to use high magnetisation ferromag-netic layers. Early devices used the high magnetisation $Fe_{40}Co_{60}$ alloy for the pinned layer whose value of M_s exceeds 20 kG. $Ni_{80}Fe_{20}$ (permalloy) was used as the free layer with M_s of the order of 20% that of CoFe. $Co_{40}Fe_{60}$ in bulk has a saturation magnetisation of 24 kG which is the highest value known. However it also has a moderate anisotropy resulting in a coercivity of ~100 Oe. This value is too high to enable its use in a GMR sensor. Initially by using a multilayer and subsequently by control of grain size below the exchange length, $R_0 = \sqrt{A}/M_s$ where A is the exchange stiffness, this problem was overcome.

Figure 5.6(a) shows a typical structure of a spin-valve device and Fig. 5.6(b) shows the magnetic structure from the early years of this century. However modern spin-valve sensors for hard disk drives where data is now oriented perpendicular to the plane, have the sensor aligned such that the plane of the layers in the stack lie perpendicular to the plane of the disk. This has a number of major advantages not least that there is no need to lithographically define the sensor to the track width and the on-track linear density in such an orientation is controlled by the thickness of the layers in the stack. This is discussed further in Section 7.5.

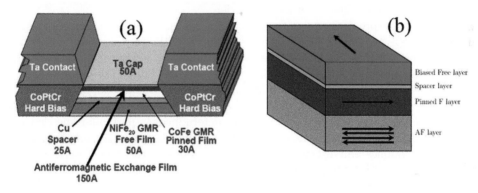

Fig. 5.6: (a) An early spin-valve structure (Thompson, 2008) and (b) the basic magnetic structure schematic.

Following the introduction of the GMR sensor in commercial production which occurred in 1997, the already rapid growth in the aerial density that could be stored on a disk doubled in terms of the growth rate over the next decade. One often hears mention of the famous Moore's Law of the decrease in size in devices on a silicon chip but in this era the rate of growth of aerial density on hard disk drives far exceeded anything achieved by the semi-conductor industry. Hence the discovery of GMR and the subsequent development of the spin-valve sensor could be said to have been the most important development in magnetics technology since the discovery and development of permanent magnets based on the alloy NdFeB back in the 1980s.

5.4 Tunnelling Magneto-Resistance (TMR)

The first observation of TMR was in 1972 on granular $Ni-SiO_2$ systems (Gittleman *et al.*, 1972). The first observation in a thin film structure was made by Jullière (1975). However both these measurements were undertaken at very low temperatures and so their importance was not realised until the discovery of GMR discussed in the previous section. Also the development of systems for thin film growth were required for the major subsequent advances to be made. This occurred in 1995 when Moodera *et al.* published room temperature TMR results of about 12% in a thin film structure consisting of $CoFe/Al_2O_3/Co$ where the Al_2O_3 formed an amorphous insulating tunnel barrier. Significant effort both in academia and industry went on to develop devices based on such amorphous tunnel barriers and TMR values of about 50% were observed in patterned devices at room temperature.

Not only did the discovery of GMR stimulate a renewed interest in TMR among experimentalists but it also stimulated an enormous interest from those involved in theoretical modelling in particular with regard to the detailed electronic band structure and the effect of an insulator being in close proximity to the metallic layers. Predictions of potential TMR values in excess of 5000% were made primarily by the group of W. H. Butler at the University of Alabama for the case were MgO was the barrier (e.g. Butler *et al.*, 2001). Of course these predictions related to single crystal structures with an epitaxial crystalline layer as the barrier.

Following these predictions significant experimental effort resulted and whilst TMR values of 1000s% were not observed, values of 200% TMR were measured (Parkin *et al.*, 2004). However these again were observations made on systems consisting of single crystals. In many respects the development and understanding of the processes giving rise to these very large values of TMR were supported by the development of atomic or even sub atomic transmission electron microscopy techniques which enabled the precise degree of epitaxy in such systems to be determined. In particular knowledge of the precise atomic structure of the epitaxial growth and the use of advanced computer modelling techniques enabled quite precise calculations of the tunnelling density of states and in particular whether it was the majority or the minority spin conductance that dominated the system.

The structure of a TMR stack is almost exactly the same as that for a GMR device with the exception that the barrier between the two magnetic layers is now an insulator as shown in Fig. 5.7. At this time it is usually MgO which is crystalline and grown epitaxially on the Fe based magnetic pinned layer. Because it is a spatial effect and conduction via tunnelling occurs, a similar oscillatory coupling occurs. Originally GMR and TMR stacks were operated from the second AF peak with the spacer layer up to 2.5 nm thick. However due to improvements in growth techniques the most advanced devices produced by the hard drive industry operate from the first peak and are only 0.9 nm thick. For MgO this is only 3 atoms thick. Hence magnetics technology now operates in the realm of atomic engineering.

In the field of TMR, $\Delta R/R$ is usually normalised to R parallel (R_P), i.e.

$$\frac{\Delta R}{R} = \frac{R_{AP} - R_P}{R_P} \tag{5.7}$$

and again the resistance is the sum of the majority and minority channels. Assuming that there is no spin mixing during tunnelling then

$$\frac{\Delta R}{R} = \frac{2P_1P_2}{1 - P_1P_2} \tag{5.8}$$

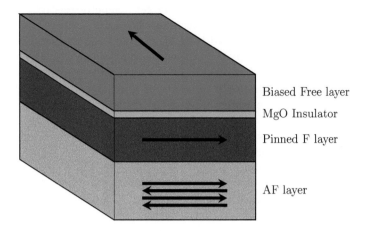

Fig. 5.7: Schematic of a TMR stack showing the insulating MgO barrier (green).

where P_{1-2} are the spin polarisation of the conduction electrons in the two channels. The spin polarisation is a result of the density of states at the Fermi level. Hence,

$$P = \frac{f(\Delta EdE)_+ - f(\Delta EdE)_-}{f(\Delta EdE)_+ + f(\Delta EdE)_-} \qquad (5.9)$$

Fairly rapidly and with the enormous capability of the hard drive industry, initial TMR read heads were developed first by Seagate Inc in Minneapolis who launched the first commercial disk drive containing a TMR read head. The major problem with such heads was that the sputtered semi-amorphous Al_2O_3 layer was subject to the development of small voids or pinholes because of the requirement that the barrier was less than 10 nm thick. This resulted in relatively low yields in production systems compared with GMR based heads. However in fairly short order and following the calculations from the group of Butler, it was realised that MgO would perform in a much better way as it could be grown in crystalline form.

This was achieved and patented by the Canon Anelva company of Japan who produced a process whereby an underlayer of CoFeB was first deposited by DC magnetron sputtering and exchange biased to an antiferromagnetic layer in a similar manner to that used in a GMR device. This CoFeB layer deposits in an amorphous form and was followed by the growth of an amorphous MgO layer above it by rf sputtering with a high magnetisation CoFe layer constituting the free layer. Under ordinary circumstances it would take a temperature approaching 1000 K to crystallise the MgO. However the CoFeB layer was able to be crystallised at a much lower temperature of around 600 K and as this crystallisation occurred it stimulated the MgO layer itself to crystallise and more importantly to crystallise in a manner in a granular film that was highly epitaxial with the CoFeB. Following patterning this was able to produce devices based on TMR that had a significantly larger magneto-resistive coefficient and thereby able to detect smaller bits to higher resolution.

Whilst the advent of the TMR read head allowed for even further advances in the process of data density in hard drives it is the case that at the time the on-track and track width density in a hard drive was limited by the lithographic process because the layers in the stack were parallel to the plane of the disk and hence it was the size of the element that could be reliably produced that was the limitation in the advance of data density.

However it was soon realised and indeed had been known for a long time, that the potential data density would be far greater if the data was stored in bits oriented perpendicular to the plane as discussed in Section 7.3 *et seq*. In this case the read head is required to be oriented with its layers perpendicular to the plane of the disk. Hence the on-track data density would be controlled by the thickness of the layers in the head rather than by a lithographic process. The current now has to flow perpendicular to the plane of the layers in the stack, i.e. a CPP geometry, which is much simpler to fabricate.

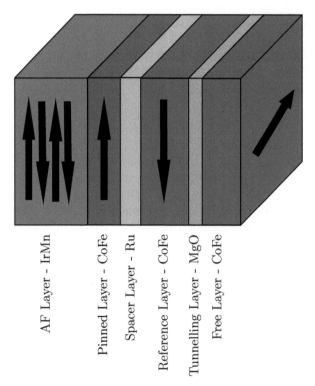

Fig. 5.8: Schematic of a typical read head stack containing a TMR junction.

Therefore from the early 1990s a major effort was mounted by the hard drive companies and notably Seagate to develop a hard drive where the data was stored in that orientation. Originally it was anticipated that such a major paradigm shift could be achieved within a two year period but a further complication occurred when it was realised that the signal profile in this orientation was radically different to that for the in plane case but none-the-less by the mid to late 1990s the paradigm shift was achieved allowing a further surge in the data capacity of a standard disk. Again this is discussed in Section 7.3 *et seq.*

To this end a number of technological and scientific advances also promoted the growth in on-track data density. These included the York Model of Exchange Bias discussed in Section 4.4, where the concept that smaller grains were always better was overthrown because of the thermal activation processes involved. It became clear and was implemented by the major data storage companies, that because of the critical influence of the grain volume in the antiferromagnetic layer, it was preferable to have grains with a larger lateral grain size and a reduced thickness. This is because previously the antiferromagnetic layer was the thickest layer in the stack and any reduction would therefore enhance on-track data density.

Figure 5.8 shows a schematic of a typical read head stack containing a TMR junction for use to read date oriented perpendicular to the plane of the disk. It should be noted that an extra feature now in common use is to grow a synthetic antiferromagnetic layer as discussed in Section 7.5 to stabilise the pinned layer and enable the TMR to be generated by two high moment CoFe ferromagnetic layers.

5.5 Spin Transfer Torque (STT)

As discussed in Chapter 8, for the last 20 years or more it has been a long held dream to create a solid state magnetic memory. This was based on the discovery of GMR and a matrix of storage cells and what were known as bit lines and word lines were required to both switch the element and to read the resulting GMR. However it had long been recognised that as well as the obvious difficulties in fabrication, there were also going to be significant problems with the level of current required to generate a conventional magnetic field to switch an element without its energy barrier being so small that it would be subject to significant thermal instability discussed in Chapter 3.

Despite these difficulties MRAM technology has been available for specialist applications for many years. In particular military systems have used MRAM technology because the competitive DRAM or SRAM technologies would not be radiation hard. In this context this does not refer to nuclear radiation but to electromagnetic radiation that can result from a nearby explosive blast. Recall that whilst we see only the blast in the optical region of the spectrum there is a wide range of frequencies generated which are capable of wiping Flash Memory. Of course in the military environment energy consumption was not an issue but such systems were unsuitable for mass consumer use.

As we have seen, a non-polarised electric current will become polarised either when that current passes through a relatively thick layer of a magnetised material but more importantly polarisation of the spins of the conduction electrons occurs when they pass through a spin tunnel junction which acts like a spin filter. In 1996 Luc Berger and John Slonczewski of IBM working to some extent in collaboration realised that if one had a spin polarised current such that the electron spins were oriented in one direction and were then passed into a small single domain element, the electrons would become reoriented into the direction of the magnetised element. By a simple consideration of the conservation of angular momentum it is then clear that such a spin polarised current would exert a torque on the magnetisation direction of the moment of the element. When correctly engineered this spin torque or as it is now called, spin transfer torque (STT), would be capable of switching the magnetisation without the flow of a normal current.

In a seminal work John Slonczewski (1996) undertook calculations which predicted this effect and subsequently Katine in 2000 and co-workers produced small elements and demonstrated that such effects could be realised. Figure 5.9 shows the structure studied by Katine *et al.* and the resulting switching of nanopillars which were 130 nm in diameter and constructed by electron beam lithography, evaporation, lift off and ion beam milling (Katine *et al.*, 2000).

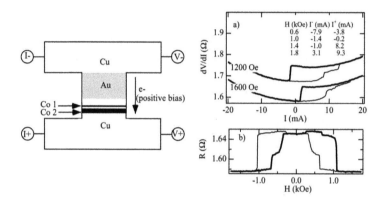

Fig. 5.9: Schematic of an STT structure and corresponding switching characteristics. Reprinted figure with permission from [Katine *et al.*, Physical Review Letters, 84 3149, 2000]. Copyright (2000) by the American Physical Society.Electronic format permissions: 10.1103/PhysRevLett.84.3149

This remarkable effect has now led to the development of magnetic random access memory or MRAM technology. At the time of writing, MRAM chips are in partial production in companies such as Global Foundries in Singapore, Toshiba in Japan and a number of companies in the United States. As discussed in Chapter 8 one of the most critical features of spin transfer torque which again is based on MgO based tunnel junctions, is that such structures can be grown directly on and integrated with CMOS silicon chips (Dieny *et al.*, 2020). This has major implications for production cost due to the lack of interconnects and also for the reduction in energy consumption in computer based devices. At the present time the energy consumption in data storage and in particular in cloud computing, makes an enormous contribution to CO_2 emissions and hence global warming.

The other major advantage for computer technology of the fact that we now have STT-MRAM is that in principle, a magnetic element can switch a thousand times faster than a silicon element but in practice a factor of closer to 100 is realised. Of course since no current is required to flow then energy consumption to switch the magnetic element is reduced by almost an order or magnitude. At the time of writing the renowned group of Hideo Ohno at Tohoku University in Sendai, Japan have demonstrated real scale chips albeit not of high density, which contain MRAM cells with a node size of 8 nm (Watanabe *et al.*, 2018) and the group of Bernard Dieny at Spintec in Grenoble have fabricated nodes as small as 4 nm on a demonstrator scale (Almeida *et al.*, 2022). Again this is due to the advances in the fundamental physics that underpins magnetics technology and notably the work of Slonczewski, Berger and Katine. In particular the advances in thin film deposition technology and advanced lithographic systems, which enable such small devices to be produced on an economic scale.

Part II

Applications of Magnetic Nanoparticles and Granular Thin Films

6

Ferrofluids

6.1 Definition and Preparation

Ferrofluids are ultra-stable dispersions of magnetic nanoparticles in a wide range of carrier liquids. The particles must be chemically stable, stable against gravitational sedimentation and also stable in a magnetic field gradient. To achieve the ultrastability required, magnetic nanoparticles, typically 10 nm in diameter or less are dispersed with the use of one or more surfactants which are long chain molecules with an active radical such as carboxylic acid $(COOH)^-$ which bonds readily to the surface of metallic and oxide particles, with the hydrocarbon or fluorocarbon chain extending out into the liquid. These surfactants induce what is known as a steric or entropic repulsion because for the particles to approach each other closely requires the surfactant molecules to be ordered in some way leading to an entropic repulsive force. The nature of the entropic repulsion is shown in Fig. 6.1.

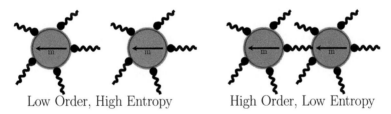

Low Order, High Entropy   High Order, Low Entropy

Fig. 6.1: Steric, or entropic, repulsion of two magnetic particles, where in order for particles to approach each other, an increase in order must be achieved and so energy added to the system.

Because of the oxidation problem, almost all commercial ferrofluids used in applications are made from nanoparticles of magnetite (Fe_3O_4) although often over time such particles oxidise through to γ-Fe_2O_3. This tends to induce a reduction in the overall magnetisation of a fluid of the order of 20% but it should be noted that in fine particle form both magnetite and maghemite have a lower magnetisation than the bulk of such materials due to both internal crystallographic disorder and the misplacing of ions within the crystal lattice.

As discussed in Section 2.7 the magnetic behaviour of magnetite is via superexchange and the distribution of Fe^{2+} and Fe^{3+} between tetrahedral and octahedral sites. When nanoparticles are produced this distribution is almost always non-ideal. In consequence the particles produced as described below, generally have a saturation magnetisation of 70-75 emu/g compared with the bulk value of 92 emu/g. Also a large fraction of the atoms are in surface layers and are hence not ordered crystallographically.

A wide range of carrier liquids can be used but because there are no known surfactants, this excludes liquid metals such as mercury. Ferrofluids can be prepared in water, a wide range of hydrocarbon oils and increasingly in perflourinated polyether oils (PFPEs) which for vacuum applications, can have lower vapour pressures at room temperature and are also resistive to corrosive environments.

It is relatively simple to prepare a ferrofluid suitable for demonstration purposes by the well known coprecipitation method (Khalafalla and Reimers, 1980). In this process a solution of Fe^{2+} and Fe^{3+} are first prepared in the approximate ionic ratio of 1:2, however in practice a slightly lower ratio is used because some of the Fe^{2+} ions oxidise through to Fe^{3+} during the preparation process. Typically iron chlorides are used as both are readily soluble in water and are inexpensive. The solutions are then added rapidly to an alkali solution typically of sodium hydroxide (NaOH) with rapid stirring. This process leads to the nucleation of very small particles of Fe_3O_4 via the precipitation of the two oxihydroxides which dehydrate in the alkaline solution. The solution is stirred with moderate heating for a period typically between 30 minutes and one hour to allow for particle growth to occur via aggregation of the smaller particles. The moderate heating tends to induce improved crystallinity of the resulting nanoparticles. Fortunately such a process naturally produces particles with typical median diameters in the range 8 to 12 nm. Such particles are always single domain as discussed in Section 2.4 *et seq.* The chemical reaction is shown in eqn 6.1

$$FeCl_2 + 2FeCl_3 + 8NaOH \rightarrow Fe_3O_4 \downarrow + 8NaCl + 4H_2O \qquad (6.1)$$

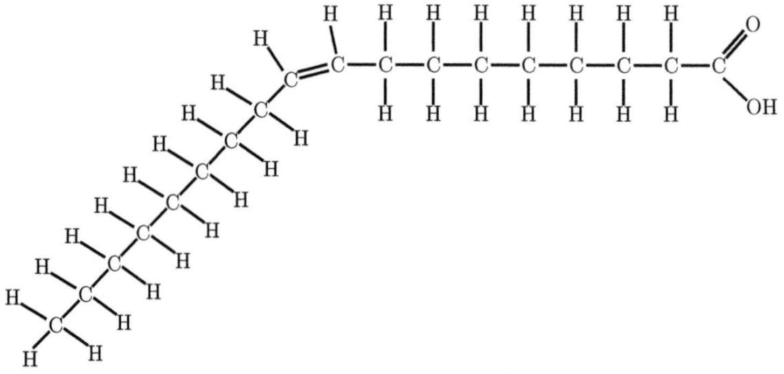

Fig. 6.2: Schematic diagram of the structure of oleic acid, showing its non-linear carbon chain.

In order to disperse the particles thus produced into a suitable hydrocarbon such as a paraffin oil with a moderate vapour pressure or a solvent such as hexane, it is necessary to coat the particles with a suitable surfactant. The most commonly used surfactant is oleic acid. Oleic acid is relatively inexpensive and has a non-linear hydrocarbon chain leading to a carboxyl group at its termination. Figure 6.2 shows the chemical structure of oleic acid. The carboxyl group readily bonds chemically to the surface of the magnetite nanoparticles and the presence of the non-linear chain leads to enhanced steric repulsion between the particles. Of course oleic acid is not directly soluble in water but is readily solvated in an alkali solution with a pH of around 9. Again the preparation is stirred relatively quickly and generally heated to almost the boiling point of the solution to promote the chemical bonding and to promote the separation of any aggregated particles that have formed prior to coating. For stability the hydrodynamic size of the particles, i.e. particle + surfactant, must be similar to the length of the molecules in the carrier oil as shown schematically in Fig. 6.3.

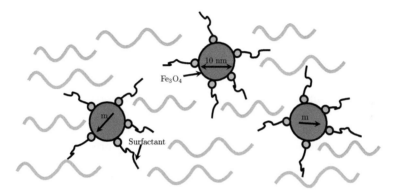

Fig. 6.3: Schematic diagram of the structure of a ferrofluid.

Due to the high solubility of oleic acid in an alkali solution it follows that as an oil based material, it is insoluble in low pH even at the pH neutral value of 7. Hence if the resulting solution is gently acidified with moderate warming using any dilute acid such as HCl the oleic acid becomes insoluble and as the pH becomes mildly acid typically pH 4 to 5, the coated particles become highly hydrophobic. This process is clearly visible in a glass beaker because the coated particles form what we refer to as a clag, which is basically an oily sponge like material containing the surfactant coated particles, which separates completely leaving a clear solution containing the residual salts. The separation is so complete that it is then possible to remove the 10 nm coated particles from the solution using something like a culinary sieve used for sieving flour. It is for this reason of removing the particles from the preparation vessel that carbocyclic acids which are generally insolvent in acidic or neutral solution, are used.

The resulting coated particles then need significant washing to remove the residual chloride irons and others that are trapped within the clag. This is achieved by washing the particles repeatedly preferably in water close to its boiling point. To achieve the removal of the ions from this sticky material requires as many as 6 or even 10 washes. Finally the particles are washed perhaps 2 or 3 times in acetone which facilitates the removal of any excess and unbonded oleic acid as well as removing much of the water.

Preferably the coated particles are not allowed to dry out but are added in the appropriate quantities to the final carrier liquid. While solvents such as heptane or toluene will readily disperse the coated particles for demonstration purposes, it is best to use a somewhat longer chain paraffin oil and perhaps the oil of choice is the isoparaffin oil Isopar-M, which is a product of Exxon corporation (https://www.exxonmobilchemical .com). Fluids dispersed in Isopar-M will remain liquid when exposed to normal room temperature for perhaps a week or two without the paraffin oil evaporating. In the same series from the same manufacturer, longer chain paraffin oils are also available which have lower vapour pressures and hence do not dry out readily. However an Isopar-M fluid contained within a bottle with a suitable cap seal will not dry out for many years. To disperse the particles into the final oil a measure of the clag is added to the appropriate quantity of the oil and the resulting mixture is then heated to drive off any residual acetone or entrapped water within the clag. It should be noted that this procedure and preferably the entire process from the start, should be undertaken in a suitable fume cupboard.

Because of the compatibility of oleic acid with such isoparaffin oils, the particles readily disperse with moderate heating. This process, depending on the volume and the temperature used, can take up to an hour. The resulting fluid should be fully stable not only against gravitational precipitation but also in the presence of a magnetic field or even a relatively strong magnetic field gradient. Enhanced stability of the resulting ferrofluid can be achieved by centrifugation which can remove any large aggregates of undispersed material.

Because the hydrodynamic size of the particles is similar in scale to the length of the molecule of the isoparaffin oil the resulting ferrofluid appears and indeed behaves as a truly homogenous magnetic liquid. Because the fluid appears as a uniform magnetic liquid a number of unexpected phenomenon can be observed.

It should be noted that for the application of ferrofluids very often quite exotic carrier oils are used. For example to have a ferrofluid for use in a vacuum seal as discussed in Section 6.4.1 it is necessary to use exotic hydrocarbon oils which can have vapour pressures as low as 10^{-11} mbar. For the case of PFPE oils required by the semiconductor industries, fluids with vapour pressures as low as 10^{-16} mbar are also available. For applications in loudspeakers, discussed in Section 6.3, the particles are dispersed in organic liquids which have to be compatible with the glues used in the manufacture of the loudspeaker but also must resist the very high temperatures that can occur particularly in large speakers and also in some smaller speakers where a significant level of power is required, such as those found in televisions and in car audio systems.

For a number of applications it is necessary or desirable to produce a ferrofluid in water. This is readily achieved by substituting water for the iso-paraffin oil and then adding a secondary surfactant. Two secondary surfactants are found to be very effective, the first of which is achieved by adding a chemical known as triethanolamine which transforms outer layers of oleic acid already on the particle, into a water compatible form by the formation by triethanolamine oleate. However there are many surfactants which are capable of dispersing oil based materials into water, the most commonly used of which is sodium dodecyl benzene sulphonate (SDBS) which is simply added to the water prior to the addition of the oleic acid coated magnetite. Again gentle to moderate warming achieves the dispersion. Water-based colloids of this type are generally less stable than oil-based materials. These water-based fluids find application primarily in sink float magnetic separation also described in Section 6.4.3.

The typical ferrofluids produced as described above lend themselves readily to analysis by transmission electron microscopy. Figure 6.4(a) shows a typical micrograph for a colloid based on the technique described above.

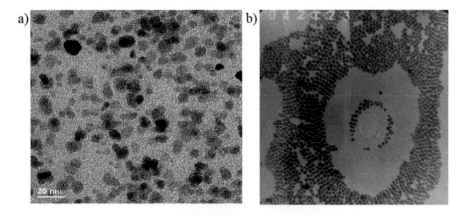

Fig. 6.4: TEM micrograph of typical (a) Fe_3O_4 and (b) Co nanoparticles.

Figure 6.4(b) shows a ferrofluid containing cobalt nanoparticles prepared by the decomposition of dicobalt octacarbonyl $(Co_2(CO)_8)$ in toluene which is refluxed, where oleic acid was dissolved in the toluene prior to the process. With the reflux of the toluene this compound releases significant quantities of carbon monoxide and relatively well crystallised particles of cobalt in an fcc phase are produced. The coating with oleic acid is now much improved due to the fact that the surfactant is incorporated into the particles themselves prior to them becoming magnetic. It also inhibits the growth of the particles thereby producing a very narrow size distribution. Because of the presence of refluxing toluene and the emission of carbon monoxide this process must always be undertaken in a suitable fume cupboard. Unfortunately as mentioned above, the resulting cobalt particles are not stable against oxidation. For the case of magnetite shown in Fig. 6.4(a) it will be noticed that there is a certain degree of aggregation of the particles. This occurs because prior to adding the dispersant, larger particles will seed the aggregation usually with smaller particles due to the dipole-dipole interaction discussed in Section 3.8.1 and also due to Van der Waals forces which, once the aggregates begin to form, almost completely inhibits the full dispersion of the particles.

6.2 Unusual Phenomena in Ferrofluids

6.2.1 Surface Instability

In normal everyday life we are used to seeing liquids either contained in some kind of vessel be it large or small or even seeing a liquid on the surface of a lake. In the absence of any type of mechanically or weather induced turbulence the surface of the liquid generally appears flat with the exception that when a liquid is contained within a vessel, some upturning of the surface is seen near the walls of the vessel due to surface tension effects. The reason why the surface of a liquid appears flat when in a vessel is that over the length scale of a typical vessel such as a laboratory beaker, the lines of force due to gravity are essentially parallel and therefore there is a uniform flat surface on the liquid. In the case of a ferrofluid the effects of gravity are identical. However it is now possible to add an additional force to a ferrofluid which is not the case for any other liquid. For example the ferrofluid can be exposed to magnetic field gradient either across the surface or through the depth of the sample. The question then arises of what is the equilibrium form of the surface when a strongly varying magnetic field gradient is applied?

Fig. 6.5: Surface instability in a ferrofluid sample.

The rather startling result is shown in Fig. 6.5. As can be seen in the figure the surface breaks up into a series of spikes, the height of which is determined by the magnitude of the field gradient and the strength of the ferrofluid. For this effect to be observed a relatively strong ferrofluid with a magnetisation of about 400 Gauss ($=4\pi M_s$ (emu/cc)) and exposure of a thin layer of up to 5 mm thick to the surface of a strong, usually rare earth, permanent magnet. It has been stated that the nature of the spikes tends to follow the lines of flux coming from the surface of the permanent magnet which is generally true but the diameter of the spikes in no way represent the lines of flux. What is true is that spikes when viewed from above do tend to lie in an hexagonal array. This of course is the equilibrium position for dipoles acting under a repulsive force due to them all being aligned in the same direction. This phenomena is known as the *"Surface Instability"*. Note that the spikes remain liquid and the particles do not pull out of suspension.

6.2.2 Labyrinthine Instability

A second even more strange phenomenon occurs if two glass plates are used with suitable spacers with a typical thickness of a couple of millimetres to form a very narrow tank preferably with dimensions between 5 and 10 cm. If the tank is filled approximately half full with ferrofluid then subsequently water can be very carefully syringed into the tank and will float on the surface of the fluid. If a magnetic field is now applied perpendicular to the narrow dimension of the tank, then over a period of a few minutes it is found that strands of the ferrofluid penetrate down into the water and then progressively sub-divide to form a labyrinth. Similarly of course, strands of the water also form within the ferrofluid resulting in two complex labyrinths in the structure of the two liquids.

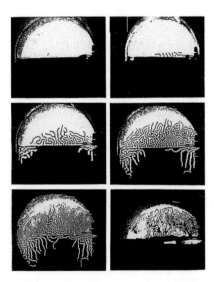

Fig. 6.6: Labyrinthine instability in a ferrofluid sample.

There is no simple physical explanation of this phenomenon which is known as the Labyrinthine Instability. Thankfully both the surface instability and the labyrinthine instability have been beautifully explained in complex mathematical terms by R. E. Rosensweig in his book "*Ferrohydrodynamics*" (1985). The "*Labyrinthine Instability*" is shown in Fig. 6.6 where the cell was placed between a pair of coils formed by taking the pole pieces from a small laboratory magnet and the image was taken using a camera with a bright light at the other side of the magnet so that the instability could be viewed through the coils. The first five images were taken over a period of 45 minutes with the sixth being taken when the field was turned off.

6.2.3 Non-magnetic Alignment

Generally non-magnetic entities and the entirety of space experiences only relatively mild magnetic fields generated by the motion of iron for example in the centre of some planets. Using a ferrofluid it is possible to put a non-magnetic entity into a relatively strong magnetic environment created by an applied field and the magnetisation of the fluid. Because of the fluid matrix, movement of a non-magnetic entity is possible.

Consider now a microscope slide onto which is placed a polystyrene sphere of size ~20 μm which is then immersed in a ferrofluid and obviously a coverslip applied on the top. If the system is now exposed to a magnetic field in the plane of the slide then flux lines coming from the north pole of the magnet used to generate the magnetic field, will terminate at the surface of the polystyrene sphere. Similarly because the polystyrene sphere has almost zero magnetic permeability the flux line will re-emerge at the other side of the polystyrene sphere whilst not transmitting through the sphere itself and will then progress to the south pole. However where a flux line from a north pole is terminated, that generates an induced south magnetic pole on the surface and likewise where the field line appears on the opposite side of the polystyrene sphere a north pole is generated. Hence the non-magnetic polystyrene sphere now has an induced magnetic moment. This situation is shown in Fig. 6.7. Of course any other non-magnetic body such as a glass sphere or a metallic sphere would experience the same phenomenon. The moment induced is simply given by the susceptibility of the ferrofluid multiplied by the applied field multiplied by the volume of the sphere as shown in eqn 6.2.

$$\mathbf{m}_{sphere} = \chi \mathbf{H} V \qquad (6.2)$$

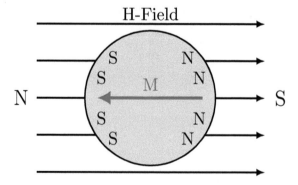

Fig. 6.7: Induced moment on a polystyrene sphere.

In a way this is analogous to Archimedes Principle as the induced moment on the sphere is equal to the moment of the fluid it displaced. Polymer spheres are readily available commercially and can be purchased in sizes ranging from about 0.2 μm up to around 50 μm. Such spheres are manufactured to be highly uniform in size and are very often used for calibrating electron and even optical microscopes. It should be noted that here we are not referring to magnetic polymer microspheres which are discussed in Section 6.6. If we now take a similar microscope slide again with a drop of ferrofluid and a drop of polymer spheres with a diameter of 20 μm then each sphere will acquire an induced moment given by eqn 6.2. As indicated above the net moment on such spheres is really quite large so in the presence of a magnetic field and with the strength of the induced moments, such polymer spheres interact and thereby in the presence of a field in the plane of the microscope slide, line up to form long chains. This is shown in Fig. 6.8(a).

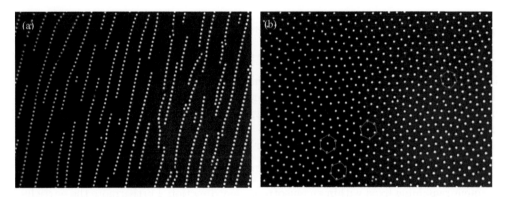

Fig. 6.8: (a) Chain ordering of polymer spheres in the direction of the applied in plane field and (b) hexagonal arrangement when the field is applied perpendicular to the plane of the microscope slide.

If a field is now applied perpendicular to the plane of the microscope slide again induced moments following eqn 6.2 are generated on the polymer spheres. However due to the presence of the cover slip, the spheres cannot form chains along the direction of the field lines and hence since all the moments are aligned in the same direction the polymer spheres now repel each other. Assuming the viscosity of the fluid used is relatively low and after waiting for several minutes, the polymer spheres then become ordered forming a hexagonal or a triangular lattice as shown in Fig. 6.8(b). It should be noticed that the two images shown in Fig. 6.8 were taken in our laboratories as part of a public lecture. Indeed in the case of the chains shown in Fig. 6.8(a), simply moving a permanent magnet around by the side makes the chains of polymer spheres appear to dance. This phenomenon of the alignment of polymer spheres in a thin layer of a ferrofluid was first observed accidentally by Arne Skjeltorp of the University of Oslo (Skjeltorp, 1983).

Whilst this phenomenon of the alignment of non-magnetic entities appears to be just a simple curious occurrence, it has found application particularly in certain areas of bio-physics where for example it is required to achieve the alignment of entities like bacteria and viruses for study using techniques such as X-rays or neutron scattering. Of course if such an entity is elongated like a bacterium then all the bacteria in the sample will line up with their long axes parallel enhancing the resolution of whatever imaging technique is used.

A further interesting phenomenon occurs if for example a number of air bubbles are introduced into the liquid. In that case not only do the air bubbles line up but because of the pressure generated by the magnetic field they are observed to elongate and become needle shaped bubbles of a gas. In a further set of experiments micron scale spheres of metals were introduced into such a sample and when aligned to form chains by a magnetic field, were observed to produce a significant polariser or switch for microwave radiation (Popplewell *et al.*, 1986).

6.3   Properties of Ferrofluids

In one respect ferrofluids are completely unique magnetic systems. This is because in a liquid matrix the particles in the colloid are free to move. This freedom of movement means that it is possible for them to align with a magnetic field by two distinct mechanisms. The first is the Néel reversal process discussed in Section 3.1 whereby the magnetic moment is able to switch over an anisotropy energy barrier. The second mechanism arises because a particle having a single relatively large magnetic moment compared to a paramagnet can also induce a physical rotation of the particle with the moment remaining fixed within an anisotropy easy direction. Furthermore the particles are free to move and in the presence of a DC magnetic field, they can agglomerate together to form long chains of particles which may then subsequently rotate.

In addition, ferrofluids prepared by methods such as that described in Section 6.1 will in general contain a significant number of particles in aggregated structures which unless refinement techniques are used, can extend to dimensions of the order of a micron. In such aggregates the moments of the particles will rotate or the particles themselves will rotate so as to form flux closure configurations. In such configurations the overall moment will probably not be zero but will be substantially reduced compared with the expected moment for that number of particles.

Such agglomerates will then have their own micromagnetic stability and it is found that they in general will align with a magnetic field by rotation. In should be noted that even in the presence of a solenoidal AC field similar agglomerations and rotation effects can occur for, whilst the time averaged magnetic field in a given direction may well be zero, there will always remain an RMS value of the applied field which will result in behaviour in a manner very similar to a DC magnetic field, albeit with a reduced amplitude. These effects lead to some perhaps unexpected and certainly unusual types of behaviour but for simplicity we will consider the case in the first instance where the degree of aggregation is very limited. Such a system can be produced by e.g. centrifugation of the colloid preferably in a low viscosity medium, so that large aggregates on the micron scale are removed.

6.3.1   Ideal Superparamagnetism

As discussed in Section 3.2 superparamagnetism occurs when the particles within a system are of a size below a critical diameter (D_p). This is the case where the criterion

$$KV < 25kT \tag{6.3}$$

applies where it should be recalled that the factor 25 relates to a measurement time of 100 s. This then leads to the magnetisation curve of a ferrofluid following the Langevin function.

$$M/M_s = L(\alpha) = coth\ \alpha - \frac{1}{\alpha}; \quad \alpha = \frac{mH}{kT} \quad , m = M_s V \tag{6.4}$$

where m is the moment of a particle and M_s is the saturation magnetisation of the material of which it is composed. Of course because there is a distribution of particle volumes within a ferrofluid and all other fine particle systems, more correctly the behaviour will be described via a summation of a series of Langevin functions. Figure 6.9(a) shows typical magnetisation curves measured at different temperatures for a ferrofluid produced in a very similar manner to that described in Section 6.1. As expected there is no coercivity and remanence and there is superparamagnetic behaviour confirmed by the H/T superposition of all magnetisation curves shown in Fig. 6.9(a) when plotted as magnetisation versus H/T (Williams *et al.*, 1993).

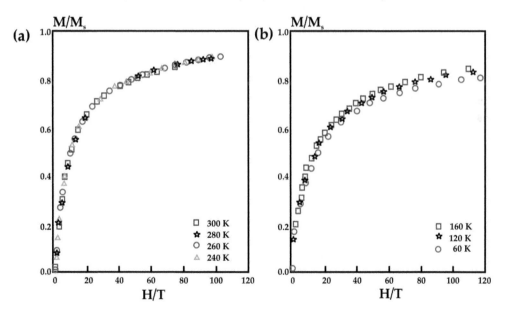

Fig. 6.9: Typical magnetisation curves for a ferrofluid measured at different temperatures showing (a) superposition and (b) non-superposition. Reprinted from Journal of Magnetism and Magnetic Materials, 122, Williams *et al.*, Superparamagnetism in fine particle dispersions, 129, Copyright (1993), with permission from Elsevier.

However while the criteria $KV < 25kT$ applies to the particles of the size found in these samples it remains the case that the anisotropy of the particles is not zero. Hence the experiment described above in the liquid state was also repeated at lower temperatures where the carrier liquid was below its freezing point. Of course at lower temperatures the value of D_p also decreased but as shown in Fig. 6.9(b) the coercivity and remanence remain zero but H/T superposition is not observed. These data serve to indicate strongly that the absence of hysteresis and exhibition of zero coercivity and remanence is only the first of two criteria that are required to produce an ideal superparamagnet. In fact given that the anisotropy is not truly zero it remains the case that a ferrofluid in the liquid state is perhaps the only true superparamagnetic system depending of course on the measurement time or perhaps the frequency of measurement as will be discussed in Section 6.5.

6.3.2   Magnetic Size Distribution

Based on the concept that the magnetisation of a ferrofluid can be described by an integral over a range of Langevin functions, Chantrell *et al.* (1978) analysed a standard magnetisation curve such as that shown in Fig. 6.9 and derived two expressions that give the median diameter of the particles in the system and the standard deviation as

$$D_{vm} = \left[\frac{18kT}{\pi M_s} \left(\frac{\chi_i}{3M_s} \frac{1}{H_0} \right)^{1/2} \right]^{1/3} \tag{6.5}$$

$$\sigma_m = \frac{1}{3} \left[ln \left(\frac{3\chi_i}{M_s/H_0} \right) \right]^{1/2} \tag{6.6}$$

where D_{vm} is the diameter of a particle having median volume, M_s is the saturation magnetisation of the fluid and χ_i its initial susceptibility. The parameter $1/H_0$ is the intercept of a tangent to a curve of M versus $1/H$ on the $M=0$ axis. Also the value of the standard deviation is that of $ln(D)$ in a similar manner to that described in Section 3.3.

The measurement of many samples over a long number of years using this technique of Chantrell *et al.* (1978) has shown that the magnetic particle size gives a consistently lower value for the median diameter than that which is obtained from studies using electron microscopy. This result which has been reproduced for hundreds of samples is further confirmation that there are magnetically dead layers near the surface of the particles due to either oxidation, crystallographic disorder or the effects of the binding of surfactants onto surface atoms. Similarly and again measured over many samples down the years, it is found that the determination of the "particle size" via X-ray crystallography to determine the crystalline size from Scherrer broadening gives a third size which is smaller than both that obtained from TEM and from the Chantrell equation. Whilst the Scherrer broadening method is known to be far from fully accurate, it does indicate that particles with a size of around 10 nm are almost never fully crystallographically ordered. This is inevitable because if one considers a particle of diameter 10 nm consisting of either magnetite or a metal such as Co then such a particle is only 30 atoms wide. Hence it is inevitable that a significant fraction of the particles would lie within say two or three atoms of the surface and would not be crystallographically coherent.

6.3.3   The Schliomis Diameter

As discussed above there are clearly two modes by which particles which are free to move in a liquid, can align with an applied DC field. These modes are each characterised by relaxation times. The first of these is the well known Néel mechanism discussed in Section 3.1 where an internal rotation of the effective moment of the particle occurs in the relaxation time now given as τ_N given in eqn 3.5.

The second relaxation time is known as the Brownian relaxation time because it involves the physical rotation of the particles. The Brownian relaxation time is given by

$$\tau_B = \frac{3V_H \eta}{kT} \tag{6.7}$$

where η is the viscosity of the carrier liquid but now it should be noted that the volume in eqn 6.7 V_H is the hydrodynamic size of the particle. This size will include the physical size of the particle but will also include the effective diameter of the surfactant coating on the surface of the particles. Additionally as the flexible surfactant molecules rotate with the particle a certain amount of the solvent in which they are dissolved will also rotate and the width of the layer of solvent which rotates may well exceed the diameter of the particle plus the surfactant coating. Hence the hydrodynamic size cannot be compared with the other sizes discussed above.

Schliomis (1974) equated the two sizes for an assumed value of viscosity $\eta = 1$ cP to obtain a critical diameter at which a transition from the Néel mechanism to the Brownian mechanism would occur. Unfortunately he used the particle physical diameter rather than the hydrodynamic diameter. None-the-less he defined the criteria for the transition as

$$V_s = \frac{kT}{3\eta f_0} exp \left[\frac{KV_s (1 - H/H_K)^2}{kT} \right] \text{ or } D_s \geq \left(\frac{24kT}{\pi K} \right)^{1/3} \text{ for } H = 0 \tag{6.8}$$

There are limitations to this very simple analysis not only in terms of the use of the hydrodynamic volume but also there is the field dependence of the energy barrier for the Néel relaxation time. In addition this discussion has assumed that a ferrofluid consists of discrete individual particles which is hardly ever the case due to aggregation processes originating in the preparation. Bogardus *et al.* (1975) observed the two rotation processes by using a pulsed applied field to magnetise the system and then observed the rate of decay of the magnetisation and found that there were clearly two distinct processes taking place.

In fact there is a very simple experiment albeit undertaken in a low magnetic field, by which the modes of magnetisation can be clearly seen. In this experiment a sample of ferrofluid is cooled to below the melting point of the carrier liquid in zero magnetic field. If the temperature is then gradually increased in a small applied field of a few tens of Oersteds i.e. below the switching field for any particles magnetising by the Néel process, then at the point where the fluid melts, if the dominant mode of magnetisation is via Brownian rotation a sudden dramatic rise in the magnetisation is observed. If the dominant mode of magnetisation is via Néel reversal a significant magnetisation is observed even in the solid state because of the absence of an energy barrier to reversal and no significant rise is observed at the melting point (O'Grady *et al.*, 1983a).

6.3.4 Pseudo Curie-Weiss Behaviour

Given that the particles in a ferrofluid behave in a manner directly analogous to that of a paramagnet, then other phenomena can occur. Obviously the particles interact via an exclusively dipole-dipole interaction but the interaction is now very strong because the moment on a single domain particle in a ferrofluid is hundreds of times larger than that on a paramagnetic ion. As we have seen because the particles are free to move then ideal superparamagnetism occurs despite the fact that the particles are strongly interacting.

In consequence measurements of the inverse susceptibility as a function of temperature show Curie-Weiss behaviour (O'Grady *et al.*, 1983b). An example of this behaviour is shown in Fig. 6.10. In this case the measurements were made on a ferrofluid containing cobalt particles having a diameter of the order of 7 nm. Because these materials were prepared via the decomposition of dicobalt octacarbonyl described briefly in Section 6.1, no primary aggregation of the particles would occur because the surfactant coats the particles prior to them becoming magnetically ordered.

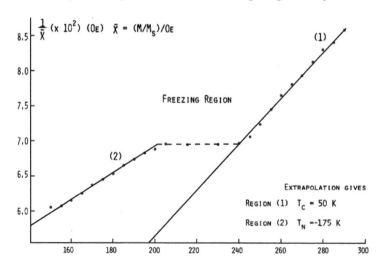

Fig. 6.10: Curie-Weiss behaviour observed in a ferrofluid containing Co particles. Reprinted from Journal of Magnetism and Magnetic Materials, 31-34, O'Grady *et al.*, Curie-Weiss behaviour in ferrofluids, 958, Copyright (1983b), with permission from Elsevier.

As can be seen in Fig. 6.10 the particles clearly exhibit Curie-Weiss behaviour and have an ordering temperature of approximately 195 K. However when measurements were made below the freezing point of the carrier liquid then once full solidification is achieved a second region of Curie-Weiss behaviour occurs. This again is highly linear but extrapolation of the data gives a Néel temperature of approximately 100 K. It is interesting to note that as the liquid freezes but does not have a well defined freezing point because of the presence of the particles and the entrained surfactants.

Of course once the material is frozen, particle movement is no longer possible and it is clear from the data that the dipole-dipole interaction becomes exclusively demagnetising and the presence of a negative intercept on the temperature axis does not indicate antiferromagnetic behaviour. In the paper the measurements shown in Fig. 6.10 were supported by Monte Carlo calculations on a system with similar parameters and similar results were obtained. This particular prediction from a Monte Carlo model is significant because to our knowledge this was perhaps the first ever large scale computer simulation of a magnetic system which was developed by R. W. Chantrell.

6.4 Applications of Ferrofluids

Because a ferrofluid behaves as a true magnetic liquid and the fact that for high quality materials the particles will not pull out of suspension even in a fairly strong magnetic field gradient, it is possible to locate a ferrofluid in a region with an appropriately designed magnetic field or field gradient and it will remain in place without the need for a container to hold it. Hence a number of applications based on this feature are possible but the commercial market for these materials is determined by two classes of device.

6.4.1 Vacuum Seals

By value the largest commercial application of ferrofluids is in the production of rotary vacuum seals that are incorporated in almost all coating systems where there is a need to open and close shutters but also in the large-scale transport systems that are used to move substrates in and out of vacuum chambers. Because of the requirement for extreme cleanliness in such coating systems, which at this time generally operate at a level of ultra-high vacuum (UHV), it is not possible to have motors or other actuators inside the vacuum chamber. The transport systems are therefore kept outside the chamber with the necessary motive power being brought in using chain drives and ferrofluid based vacuum seals. Because of the requirement to operate under UHV conditions, the ferrofluids produced for these applications must have ultra-low vapour pressures. A standard vacuum seal is usually based on a typical naphthalene oil having a vapour pressure of around 10^{-9} mbar but hydrocarbon fluids with vapour pressures as low as 10^{-11} mbar are also used. For these materials a similar preparation route to that described in Section 6.1 are used but it is necessary to incorporate further dispersants to make the oleic acid coated material compatible with the oil.

For seals to be used, particularly in semi-conductor processing and UHV, it is usually the case that oils known as perfluoropolyethers (PFPE) are used. These are exotic oils with very unusual properties for example their density is typically >2 g/cc. They also have extreme corrosion resistance as they have similar chemical structures to the commonly used plastic PTFE. This is particularly important for the etching process by which small and sometimes very small features are lithographically defined in silicon chips and devices such as Flash Memory.

For the processing of high resolution silicon based devices there is a preference to have no hydrocarbons inside the vacuum system as this can contaminate the surface of the silicon and hence the preference for the use of PFPEs. PFPEs are made with a range of vapour pressures again beginning with a standard fluid having a vapour pressure of 10^{-9} mbar but oils are available and indeed sealing fluids are available, in more exotic PFPE oils with vapour pressure as low as 10^{-16} mbar. Of course surfactants based on hydrocarbons such as oleic acid are not compatible with PFPEs but other PFPEs with carboxylic acid active radicals are commercially available which enables very similar processes to that described in Section 6.1 to be used to produce the fluids. However it should be noted that PFPE oils and surfactants are orders of magnitude more expensive than similar hydrocarbon materials and hence the use of these materials is somewhat limited except for the most demanding environments.

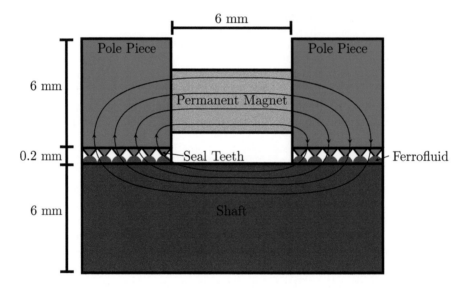

Fig. 6.11: Schematic diagram of a ferrofluid vacuum seal.

The basic design of a ferrofluid vacuum seal is shown in cross section in Fig. 6.11. The seal consists of a stainless steel barrel with flanges and bearings at either end which support a shaft. Around the shaft is positioned a permanent magnet in the form of a ring and the magnetic field is focussed via pole pieces above and below the magnet. The focussing of the field is achieved by cutting a number of grooves in the shaft or in the pole pieces so that a series of promontories appear both above and below the magnet. Typically between 5 and 10 such promontories are cut into the shaft or pole pieces either side of the magnet. When a ferrofluid is introduced into the gap it forms a series of concentric liquid O-rings which provide the seal. When the seal is pressurised either by an over-pressure or by having a vacuum on one side, the first liquid O-ring and perhaps the second and third are likely to burst but then re-seal so that there exists a series of chambers between the liquid O-rings where the pressure gradually changes. A typical ferrofluid for vacuum applications has a burst pressure of up to two atmospheres giving a 100% margin of error.

It is known that seals of this type have been run continuously for as long as 10 years without failing. However in many applications and particularly those involving the use of corrosive chemicals such as reactive ion etching, commonly used in the semi-conductor industry and in the production of read and write heads for hard drives, the fluid in the seal has to be replaced on a regular service cycle perhaps every few months. However the volume of fluid contained within the seal is relatively small due to the narrow gap of typically <0.5 mm, between the teeth in the seal and the shaft. For example a seal with a 6mm shaft typically used for shutter seals, contains only about 0.2 cc of the sealing fluid.

Even basic fluids for the production of vacuum seals are quite expensive and have a typical cost of about €10 (=$10) per cc. However the high level fluids used particularly in the semi-conductor industry can cost up to ten times this amount.

The range of seal designs that are available is quite extensive. Seals are made either with solid shafts, hollow shafts and often the seal module itself comes complete with a motor attached and where required, the seal may be water cooled if extreme temperatures are to be found in the vacuum system. For example the primary manufacturer of vacuum seals, Ferrotec Corporation, have over a hundred different seal designs available in their catalogue (https://www.ferrotec.com/).

In addition to application in vacuum seals, ferrofluid seals can also be used elsewhere for example on centrifuges where perhaps hazardous liquids or gases are to be handled and on large scale flywheels where it may be required to operate the flywheel in a chamber under reduced pressure to avoid heating. They have also been used in simple dust exclusion seals where the specification on the fluid is much lower than is the case for vacuum applications.

When ferrofluid seals were first developed back in the 1970s the industry standard specification for the fluid is that it should have a saturation magnetisation of typically 350 G–450 G (1 G = $4\pi M_s$ where M_s is measured in emu/cc) depending on the design of the magnetic circuit. The viscosity of the fluid which is a key parameter, depends critically on the nature of the oil but is typically of the order of 1000 cP (mPa·s) but again for the more exotic low pressure PFPE oils can be as high as 25,000 cP (mPa·s).

The widespread use of ferrofluid vacuum seals arises because the seals generate a perfectly hermetic seal and generate no wear on the shaft which rotates. Also there is no debris introduced into the vacuum chamber which occurs with most solid seals. The only drawback is that, particularly for the high viscosity fluids based on very low vapour pressure PFPE oils, there can be significant drag requiring a relatively high torque motor to drive the shaft. For more conventional vacuum seals it is possible to have shaft rotation speeds of up to 100,000 rpm which require water cooling, because of the viscous drag.

Whilst ferrofluid seals work well and suitable fluids are available for use not only in vacuum systems, but for example also in the containment of toxic or corrosive gases, it remains the case that ferrofluid seals generally cannot be used to seal liquids due to emulsification of the ferrofluid with the liquid which can lead to rapid failure of the seal.

6.4.2 Loudspeakers

Perhaps the biggest use of ferrofluids by volume is in loudspeakers. Almost all loud-speakers contain ferrofluids so that most readers will have some ferrofluid in their home, car and in televisions. There will invariably be ferrofluid in the loudspeakers of hi-fi systems.

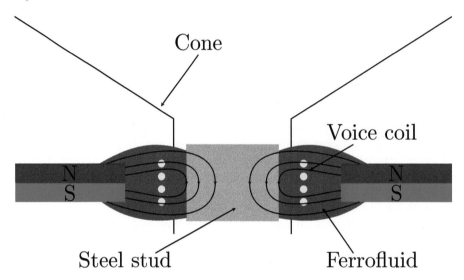

Fig. 6.12: Schematic diagram of a loudspeaker containing a ferrofluid.

The location of the ferrofluid is shown in the schematic diagram in Fig. 6.12. A loudspeaker consists usually of a paper or cardboard cone with a tube at its base around which is wrapped a coil which is excited by a variable audio frequency signal. The coil sits inside a ring magnet polarised perpendicular to the plane and usually a mild steel stud is placed in the centre of the coil to provide a magnetic circuit thus concentrating the flux in the region of the coil. A ferrofluid is introduced into the gap between the ring and the coil which both enhances and focusses the magnetic field further. It also provides three performance enhancing factors to the loudspeaker. These are as follows:

1. Because of the strength of the ferrofluid which in this case has a typical satu-ration magnetisation of between 100 G and 200 G, means that the fluid will be strongly attracted to the surface of the magnet and the steel stud. In that way the ferrofluid pushes the coil into the centre of the gap thereby avoiding any spurious development of noise by the coil touching the solid surfaces.
2. The ferrofluid also provides a conduction path for heat to be removed from the coil. In modern audio systems, particularly in cars, the output power of the system

can be as high as 20 W which would burn out the coil relatively quickly unless very thick wire was used. None-the-less the ferrofluid has to have good resistance to high temperatures as in high-powered systems the temperature of the surface of the wire can rise to in excess of 100°C.

3. The fluids used for loudspeaker applications generally have a modest to high viscosity in the range 100 cP to 5,000 cP but typically 500 cP. This serves to damp out spurious high frequencies or resonance in the voice cone itself.

The requirements for ferrofluids to be used in loudspeakers are generally met by commonly used PVC plasticisers such as dioctyl sebacate (DOS). These materials are relatively cheap and there are a wide range of such materials available enabling fluids with a wide range of viscosities to be produced. The preparation route is very similar to that described in Section 6.1.

6.4.3   Sink-float Separation

Because of the fact that a ferrofluid behaves as a true homogeneous magnetic liquid it can exhibit some rather surprising properties. Consider the case where a relatively dense object such as a small piece of copper is placed into a beaker with a conventional non-magnetic oil. The piece of copper sinks to the bottom because its volumetric density is greater than that of the oil and therefore the gravitational force which it experiences is also greater than that experienced by the oil. However once we have a homogeneous magnetic liquid it is then possible to exert an additional force on the liquid that does not change the net force on the piece of copper.

This can be done quite simply by placing a small beaker of ferrofluid above the pole pieces of a conventional laboratory magnet. Figure 6.13(a) shows this situation. In Fig. 6.13(b) when the field is applied the force on the fluid raises the copper. Given that the density of the ferrofluid is typically around 1 g/cc and the density of copper is 9 g/cc this implies that the magnetic field is generating a force on the ferrofluid which is nine times greater than that due to gravity. In this case the saturation magnetisation of the fluid is 400 G and the carrier liquid is Isopar M oil discussed in Section 6.1.

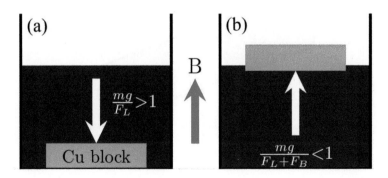

Fig. 6.13: Magnetic sink-float separation (a) in zero field and (b) when a field is applied.

In principle this technique could be used on a large scale for separation and re-cycling of a wide range of materials. However in practice the cost of the ferrofluid generally makes this preclusive. The only large-scale separation application is in the field of diamonds where there are large numbers of diamonds still to be found in spoil tips in mining areas. It is believed that certain companies are now using the sink float separation to recover these generally fairly small diamonds by having two such separators designed to sink more dense materials and then to float off the diamonds.

However recently an attempt to separate more conventional materials for recycling has been undertaken by Liquisort Recycling B.V. in the Netherlands (www.liquisort .com). They have overcome the cost issue with conventional ferrofluids by using fer-rofluids produced by large-scale grinding of much larger magnetite particles in the presence of a surfactant over long periods of time. It is believed that there are two or three such separators operating in the world at the time of writing.

6.4.4 Inclinometers

There are many niche applications which have been proposed for ferrofluids but very few if any of these are in common use. For example an inclinometer is a device that is capable of detecting angles and a ferrofluid can produce a relatively inexpensive device which achieves this.

The way a ferrofluid can produce an inclinometer arises due to the fact that if a small magnet is introduced into a vessel of ferrofluid the field will adhere strongly to the surface of the magnet and hence if a magnet is sitting in a beaker it will appear that the magnet is floating due to the magnetic forces. This induced buoyancy means that the magnet is then free to move in any direction. To achieve this a ferrofluid with a relatively modest saturation magnetisation of between 100 G and 200 G is required. A schematic diagram of such a device is shown in Fig. 6.14.

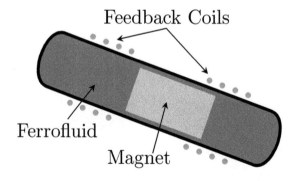

Fig. 6.14: Schematic diagram of an inclinometer device.

A small tube perhaps a few millimetres in diameter is partially filled with a ferrofluid and then a small cylindrical shaped magnet is introduced into the tube which is then sealed permanently. If two small coils are then wound around the surface of the tube and excited with a relatively low voltage applied AC field then if the magnet is closer to one end of the tube than the other then the inductance of that coil will change which can be detected very simply via a simple resonant circuit. In this way it is very easy to make an inclinometer where the magnet is maintained in the centre of the tube by applying a feedback loop from the resonant circuit to generate a DC output which will apply a small field to cause the magnet to move back to the centre. This has been done quantitatively producing a device with an accuracy of a fraction of 1°.

6.5 Magnetic Hyperthermia

Let's consider the ac hysteresis loop of a system of magnetic nanoparticles in a liquid shown in Fig. 6.15 which is for the case where the easy axes of the nanoparticles are randomly oriented and is the ideal Stoner-Wohlfarth case where all the particles are blocked (see Section 3.2) at the frequency of measurement which in this case was 47 kHz. Also the particles used were larger than those in a standard ferrofluid having a median diameter of 12.3 nm. The area within the hysteresis loop constitutes an energy loss which will be dissipated as heat. This is the principle of magnetic hyperthermia. Of course from a DC measurement or even an AC measurement at low frequencies the amount of heat generated will be relatively small and for the former case will probably be unmeasurable. However if a relatively high frequency field typically in the range 50 kHz to say 1 MHz is used then the number of times the hysteresis loop is transversed is such that a significant amount of heat is generated.

The potential for using this effect for medical applications was considered and investigated as long ago as 1957 by Gilchrist *et al.* who heated tissue samples with 20-100 nm γ-Fe_2O_3 particles exposed to an alternating field of 1.2 MHz. In the last 15 years or so there has been an enormous amount of work undertaken on the potential for this phenomenon to be used for the reduction in the size of malignant or non-malignant tumours in people. There have also been a number of human trials mainly pioneered by workers at the Charité Hospital in Berlin which built a specific piece of apparatus for these investigations and looked at the efficacy of magnetic hyperthermia on a number of different tumours (Thiesen and Jordan, 2008). For a recent review at the time of writing see Rubia-Rodriguez *et al.* (2021) and Mahmoudi *et al.* (2018).

From the hundreds of papers published on this effect there has been no real consistency in the experimental results with different researchers using particles of different types, different sizes and fields of different amplitudes and frequencies. Up until 2013 it was assumed that the heating effect could be described in terms of the susceptibility loss where at lower frequencies the moments on magnetic nanoparticles could generally follow the field. At higher frequencies the particle moments would lag behind the field direction and possibly not even move at all but when the lagging effect occurred this would constitute an energy loss. The theory of this effect was developed by Rosensweig (2002). Hence the effect would be characterised by the Néel relaxation time discussed in Section 3.1 and a Brownian relaxation time for particle rotation given by eqn 6.7.

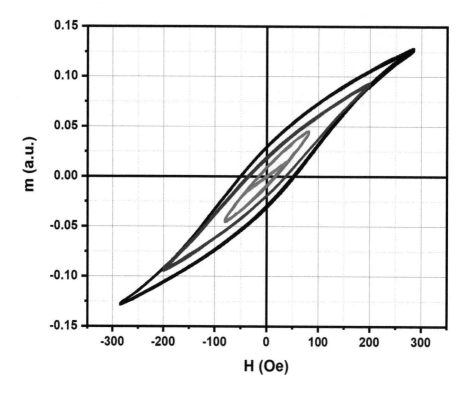

Fig. 6.15: Hysteresis loop of a sample of magnetic nanoparticles in water measured at 47 kHz and a range of field amplitudes (Clarke and Vallejo-Fernandez, 2022).

These two relaxation times can be combined to give an overall relaxation time given by

$$\tau = \frac{\tau_B \cdot \tau_N}{\tau_B + \tau_N} \tag{6.9}$$

Of course this would only apply to those particles within the system that were superparamagnetic. However at the high frequencies used, the criterion for super-paramagnetic behaviour would not be where the energy barrier was 25 kT as the measurement time is significantly shorter. In the experiments to be described here a hyperthermia measurement system was used where the frequency of operation was 111.5 kHz. At this frequency the coercivity of a particle is given by

$$H_c = H_K \left[1 - \left(\frac{54.6kT}{\pi K D^3} \right)^{1/2} \right] \tag{6.10}$$

In consequence the critical diameter $D_p(0)$=13.5 nm in zero applied field for mag-netic particles with $K = 3 \times 10^5$ ergs/cc and $M_s = 82$ emu/g. Given that ferrofluids have a typical median diameter of around 10 nm this means that a significant fraction of the particles would be blocked and would not generate any significant heating from susceptibility loss.

Furthermore most of the measurements of magnetic hyperthermia are undertaken in systems where the applied AC field is of the order of 200 Oe as was the case in the system used in our work. Hence the larger particles in the system whilst being blocked would not contribute to any hysteresis heating effect because the energy barrier to reversal is larger in field terms than the AC field applied. When this situation occurs because the particles cannot switch but are still sitting in a liquid, they will begin to rotate assuming that the Brownian relaxation time is sufficiently low to allow them to do so. In any case they will undertake some oscillatory motion which will itself generate heat due to the viscosity of the liquid in which they are suspended. Hence there are now three distinct heating mechanisms that can occur as shown in Fig. 6.16 where the parameter $D_p(H)$ above which particles will not switch is given by

$$D_p(H) = D_p(0) \left[1 - \frac{HM_s}{0.96K} \right]^{-2/3} \tag{6.11}$$

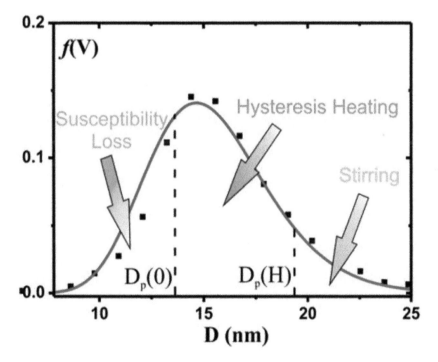

Fig. 6.16: Schematic diagram of the different contributions to magnetic hyperthermia as a function of particle size (Vallejo-Fernandez *et al.*, 2013).

Note that the factor 0.96 in eqn 6.11 applies because for the case of a system of particles with randomly oriented easy axes the value of the anisotropy field is $0.96K/M_s$

(Luborsky, 1961).

For particles with D> D_p(H) the moments are not able to switch in the usually limited field amplitude available. Under these circumstances and depending on the frequency, many particles and more importantly aggregates of smaller particles will tend to physically rotate in the liquid giving rise to viscous heating due to stirring. This effect can often be observed with the naked eye if a suitable open-top glass vessel is used. However recalling that whilst the time average field in a solenoid may be zero it remains the case that there will be a finite RMS value which will lead to the gradual alignment of these larger particles and aggregates which can account for non-linear heating rates that have been reported.

It is now clear that there are three potential mechanisms by which heat can be generated when a colloid of fine particles is exposed to a radio frequency AC field. The first of these is the susceptibility loss where the power generated is given by

$$P_{sus} = \pi f \chi'' H^2 \tag{6.12}$$

where f is the frequency of measurement and χ'' is the complex part of the AC susceptibility given by

$$\frac{\chi''}{\chi_0} = \frac{2\pi f \tau}{1 + (2\pi f \tau)^2} \tag{6.13}$$

χ_0 is the value of the initial susceptibility in a DC field and τ is the combined relaxation time from the Néel and Brownian processes given by eqn 6.9. Of course both of these relaxation times will be distributed due to the presence of a particle volume distribution and in the case of Brownian relaxation the volume being the hydrodynamic volume.

For the case of heating due to hysteresis the amount of heat generated is proportional to the frequency multiplied by the area of the loop because each loop constitutes a certain level of energy loss. It is only those particles with $D > (D_p(0))$ that contribute to the hysteresis because they are blocked at the frequency of measurement. The concept of a blocked particle in a liquid matrix may seem counterintuitive but the position arises that where the Néel relaxation time of the particle is shorter than the Brownian relaxation time and the particle will switch as if it were in a solid matrix in an attempt to lower the overall energy of the system in the shortest possible time. For such particles assuming they are randomly oriented that means that the remanence will be equal to 0.5 M_s. However it should be noted that as time progresses because of the presence of the RMS field value of the AC field along the axis of the solenoid some gradual alignment of the easy axes directions of those particles may occur. Ignoring this gradual alignment which is generally unquantifiable, the heating power due to hysteresis P_H is given by eqn 6.14. Of course it is likely that there will also be a distribution of anisotropy constants in the system and hence the parameter P_H will also be distributed reflecting the distribution of anisotropy fields that result.

$$P_H = 2M_s f \int_{V_P(0)}^{V_P(H)} H_C(V) f(V) dV \tag{6.14}$$

where f is the frequency of measurement. Unfortunately there is no known mechanism to calculate the amount of heat generated by stirring. There are formulae that can predict the generation of heat due to viscous drag for example on a sphere, but the particles in a ferrofluid are not perfectly spherical and in the case of aggregates can have quite an irregular shape which can become further elongated by the presence of the RMS field. Stirring may also lead to a lag between the alignment of the moment and the field which will create another form of heating which is analogous to that generated by susceptibility loss. We have developed a technique to measure the effect of heating by stirring which will be described.

The amount of heat generated by magnetic nanoparticles is generally quantified in terms of a parameter known as the specific absorption rate (SAR) which is given by

$$SAR = \frac{C\rho}{\phi}\frac{\Delta T}{\Delta t} \tag{6.15}$$

where C is the specific heat of the colloid, ϕ is the concentration of iron per ml of solution, ρ is the density of the colloid and $\Delta T/\Delta t$ represents the heating rate.

Table 6.1 Particle and hydrodynamic size parameters for samples dispersed in Isopar M and V (Vallejo-Fernandez *et al.*, 2013).

Solvent	Sample	D_m (nm)	σ	η (cP)	D_H (nm)	σ_H (nm*)
		±0.1	±0.01	±0.1	±1	±0.05
Isopar M	A	10.3	0.16	3.0	22	16
Isopar M	B	11.7	0.15	3.0	27	17
Isopar M	C	15.2	0.19	3.0	19	13
Isopar V	A	10.3	0.16	10.8	20	13
Isopar V	B	11.7	0.15	10.8	36	20
Isopar V	C	15.2	0.19	10.8	25	16

*Note that the distribution of hydrodynamic sizes is Gaussian and, hence, the standard deviation has units of nm unlike the standard deviation of the lognormal distribution.

All samples evaluated in this work were prepared by Liquids Research Ltd. (www.liquidsresearch.com). The particles were prepared by the co-precipitation method described in Section 6.1 followed by a controlled growth process described in Section 6.5.1 producing somewhat larger particles than those described previously, with a reasonably narrow size distribution. In principle the particles are Fe_3O_4 but in practice they will probably lie somewhere between that composition and γ-Fe_2O_3. The samples were dispersed into a hydrocarbon paraffin oil but particles from the same batches were also dispersed into a hydrocarbon wax having a melting point of around 80 °C. Two different hydrocarbon oils were used, one being Isopar-M as discussed previously, but a second isoparaffin oil having a higher viscosity was also used to examine the effects of stirring on the heating process. This particular oil was from the same family and manufacturer and is known as Isopar-V. Table 6.1 shows the key parameters for all samples including the hydrodynamic size which was determined using a photon correlation spectrometer produced by Malvern Instruments Ltd.

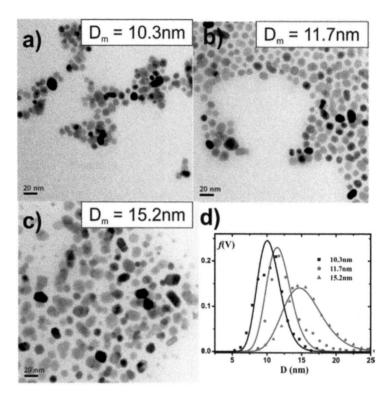

Fig. 6.17: TEM images and particle size distribution for samples used in hyperthermia studies (Vallejo-Fernandez *et al.*, 2013).

Once the particles are encapsulated in a wax this prevents stirring and enables the effect of stirring in the liquid state to be quantified approximately but also looks at the heating from susceptibility loss and hysteretic effects. TEM images of the resulting particles and size distributions measured from TEM imaging using a light box system described in Section 3.3, are shown in Figs. 6.17(a), (b), (c) and (d). Figure 6.18 shows measured values of the AC susceptibility (χ'') as a function of frequency for samples dispersed in Isopar-V. The solid lines were calculated from eqn 6.13. From this data and the data shown in Table 6.1 it is clear that the susceptibility loss does not follow the measured physical size of the particles but rather follows the hydrodynamic size shown in Table 6.1. This data in combination with eqn 6.12 shows that susceptibility loss makes only a small contribution to the overall heating power of $\sim$5% of the total and can generally be disregarded.

Figure 6.19 shows the measurement of SAR at a function of the applied field.

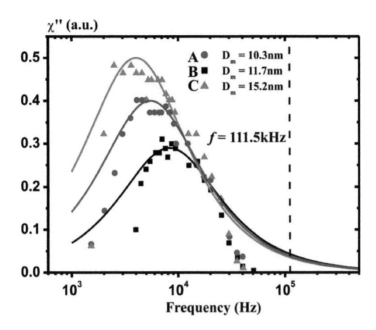

Fig. 6.18: Imaginary part of the AC susceptibility (χ") for three samples of magnetic nanoparticles (Vallejo-Fernandez *et al.*, 2013).

As expected the heating correlates strongly with the field and rises more quickly as the field is increased. This is probably due to the ability of the field to switch larger particles and aggregates which otherwise would be stirring. It is of critical importance here to look at the magnitude of the maximum value of SAR at 250 Oe. The heating rate is significantly higher in Isopar-M which has a lower viscosity than is the case in Isopar-V. This shows the effect of stirring on the overall effect. However for the particles dispersed in the wax it is clear that significantly less heating occurs with the maximum SAR being reduced by about 40%. This shows that the effect of stirring contributes significantly to the heat being generated.

This has important consequences for applications of magnetic hyperthermia for in-vivo therapies. The effective viscosity in the region where heat is required to be generated is not known and once heating occurs it is likely that damaged cells and tissues will rupture thereby changing the viscosity and perhaps lowering it substantially, which will then increase the rate of stirring and hence the rate of heat generation by that process. Hence the use of such particles in a water-based colloid will not allow for a predictable and possibly reproducible therapeutic effect.

The data from these experiments are summarised in Table 6.2. From this data it is clear that sample B has an anonymously large hydrodynamic size and a rather large SAR which correlates back to the hydrodynamic sizes shown in Table 6.1. This further indicates the difficulty of controlling magnetic hyperthermia in a liquid matrix. It is also clear that the calculated values of SAR from eqns 6.12 and 6.14 are very different

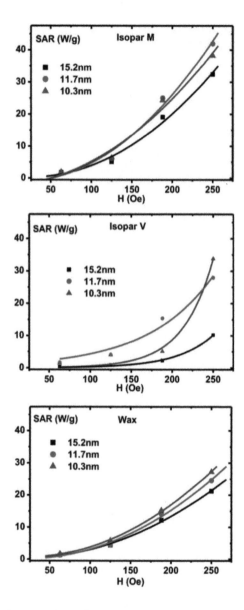

Fig. 6.19: SAR data for samples of magnetic nanoparticles dispersed in solvents of varying viscosity (Vallejo-Fernandez *et al.*, 2013).

from the experimental values and even the value when the particles are dispersed in a solid. We believe this is due to the fact that the growth process used has produced non-spherical particles and there is some evidence that in the growth process a number of particles have fused in the growth process to produce a dumbbell shape. This will lead to very large values of shape anisotropy in these particles which are not taken into account in eqns 6.12 and 6.14 which now gives values of the heating that are significantly below those measured.

Table 6.2 Experimental and calculated SAR values at a frequency of 111.5 kHz and an applied field of 250 Oe (Vallejo-Fernandez *et al.*, 2013).

D_m (nm)	SAR Isopar M (W/g)	SAR Isopar V (W/g)	SAR Wax (W/g)	Hysteretic Heating (W/g)	Susceptibility Losses (W/g)
10.3	38.2	33.8	27.2	7.1	1.4
11.7	41.9	27.9	24.5	29.5	1.7
15.2	32.4	10.2	21.2	138.3	0.7

In subsequent works (Vallejo-Fernandez and O'Grady, 2013; McGhie *et al.*, 2017; Clarke and Vallejo-Fernandez, 2022) we have undertaken experimental and theoretical studies of the role of the distribution of shape anisotropy constants which has partially resolved this issue. The distribution for a given system can be determined from the measurement of the distribution of particle elongations from TEM images (McGhie *et al.*, 2017). This can be converted easily into a distribution of anisotropy constants using eqn 2.10. The median value obtained in this way is in good agreement with the value obtained from a temperature decay of remanence measurement. However, the technique does not provide any information on the shape/width of the distribution. This is important as the width of the distribution can dramatically affect the measured SAR values for samples with identical particle size distributions (Vallejo-Fernandez and O'Grady, 2013). Given the way in which the particles are grown, their exact composition is not known exactly and in particular the value of M_s which appears as a quadratic in eqn 2.10. Hence trying to determine the distribution of anisotropy constants in real materials remains something of an unresolved challenge.

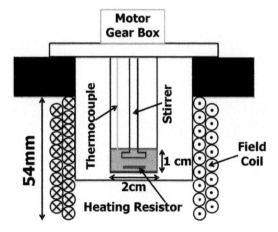

Fig. 6.20: Schematic diagram of device constructed for hyperthermia measurements (Drayton *et al.*, 2017).

A further challenge relates to the actual measurement of SAR. In all our theoretical calculations and in particular for the measurements of the heating effect, no account is taken of the heat losses to the vessel which contains the fluid. There are a number of devices that are marketed commercially (e.g. http://www.nanotherics.com) which have some issues concerning the exact field value generated which is not measured and in particular, the field uniformity. The field in the centre of a short solenoid is notoriously non-uniform and often the shape of the sample holder is quite elongated along the axis of the solenoid so that the field amplitude will not be constant. Furthermore any experiment to measure heating in a liquid, will result in convection as can readily be seen by looking inside an electric kettle as the water warms. Hence all systems should contain some kind of stirring device to ensure that the sample remains as homogeneous as possible.

Figure 6.20 shows a schematic diagram of the device we constructed including a heating resistor. As can be seen from Fig. 6.20 the sample holder now has a material contained to a depth of 1 cm inside a field coil which is 50 mm long. In this way the sample is contained in a region where the field uniformity along the axis is within 10%. The stirrer which is driven by a small motor on top of the cell ensures that the sample is homogeneous. However none of these precautions can prevent the transfer of heat from the liquid to the walls of the container. Accordingly the system was calibrated with a resistor of known value so that the sample cell could be calibrated for a rise in temperature as a function of the power generated by the resistors. This calibration curve is shown in Fig. 6.21. This experiment was undertaken using water which is the carrier liquid for particles.

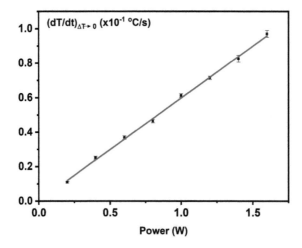

Fig. 6.21: Calibration data for hyperthermia measurements using water (Drayton *et al.*, 2017).

Subsequently the power generated by the AC field was measured by replacing the water with the same colloidal dispersions as described above. In this way the actual power per gram of Fe could be measured to a reasonable accuracy of about 2% and converted to the value of SAR. It should be emphasised that the resistors remained in place during the heating experiments with the ferrofluid because the resistors themselves will absorb some of the heat. Table 6.3 shows the corresponding results although these were in the liquid state and so the power generated includes that due to stirring.

Table 6.3 Heating calibration data (Drayton *et al*, 2017).

Sample	D_m	$(dT/dt)_{dt \to 0}$	Power	Power/gram	SAR
	(nm)	($\times 10^{-2}$ Ks^{-1})	(mW)	(W/g$_{Fe}$)	(W/g$_{Fe}$)
HyperMAG A	10.3	3.33±0.06	435±9	29.0±0.6	21.8±0.4
HyperMAG B	11.7	4.3±0.06	596±10	39.7±0.8	30.0±0.4
HyperMAG C	15.2	6.0±0.1	885±20	59.0±1.0	44.5±0.7

Alternatively the heat losses can be mitigated by taking the value of the heating power from the initial part of the curve an example of which is shown in Fig. 6.22. We found that heating with either the resistors or the magnetic fluid, the temperature as a function of time is not linear. However the heat losses to the container, the stirrer and the resistors will be at a minimum when the difference in temperature between the cell and the liquid is at a minimum since the thermal conductivity depends on the temperature difference. Hence the values of heater power that we measured are from the initial part of the curve. We believe this comparative technique and the use of a stirrer to ensure homogeneity provides the best approach to getting an accurate determination of the power generated by the magnetic nanoparticles.

6.5.1 A Dosimetric Strategy for Magnetic Hyperthermia

As discussed in detail above, one of the reasons for the slow development of therapies based on magnetic hyperthermia is the lack of consistency in the results for different materials and different samples. For example there are reports in the literature of magnetic hyperthermia measurements at frequencies ranging from 50 kHz to over 1 MHz. One reason for this lack of consistency arises from the data discussed in Section 6.5 concerning the heat generated by stirring in addition to hysteresis heating effects. The stirring effect will always be non-uniform because individual particles within a liquid environment will not only rotate in the field but also possible translate and form chains of entities during the exposure to the magnetic field. Furthermore as tissue is destroyed by the heating effect, the viscosity of the environment where the particles would be located in an in-vivo therapy would also change as previously semi solid material were liquidised. Hence there has never been a consistent approach which could inform a medical practitioner of the required dose of material required to destroy a given mass of tissue. We have recently developed a strategy that we believe

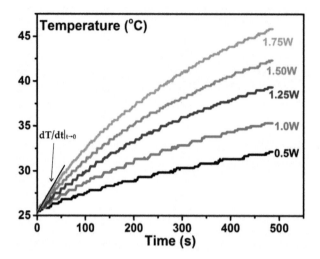

Fig. 6.22: Temperature as a function of time for different values of the heater power (Drayton *et al.*, 2017).

may overcome these problems. Referring back to eqn 6.7 the relaxation time (τ_B) for an entity to rotate in a liquid of known viscosity η is given by

$$\tau_B = \frac{3V_H\eta}{kT} \tag{6.16}$$

The variation of this Brownian relaxation time τ_B with the diameter of the particle is shown for a viscosity of 1 cP in Fig. 6.23. From this simple variation, bearing in mind that the particle volume is now the hydrodynamic volume V_H, it is clear that if a magnetic entity with a diameter of say 1 μm were to be used then a frequency in the region of 100 kHz would be sufficient to prevent any rotation of the particles. Unfortunately a magnetic entity with a diameter of around 1 μm would not be a single domain particle regardless of the material used and as such would have a very low level of magnetic hysteresis.

Referring now to Section 6.6.4 it is possible to create a magnetic entity with a size of about 1 μm as that is the typical size for magnetic polymer spheres where magnetic nanoparticles having a single domain structure are incorporated into spheres typically of a polystyrene based polymer. The generation of such polymer spheres is well established and they are often produced by the production of a ferrofluid in styrene which is subsequently emulsified in water into 1 μm droplets which are them polymerised in the normal way. We have recently examined this possibility in collaboration with scientists at Liquids Research Ltd. with whom we have collaborated for many years.

If polymer spheres with a diameter of around 1 μm incorporate standard magnetic nanoparticles produced by methods similar to that described in Section 6.1 then unfortunately the level of magnetic hysteresis resulting is very low because of a lack of crystallinity in the particles and their very small size which is typically 10 nm or less.

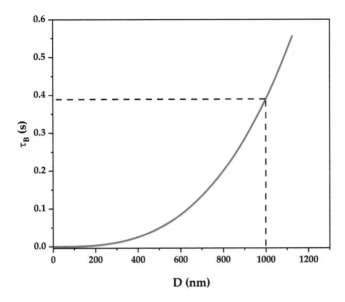

Fig. 6.23: Brownian relaxation as a function of particle size.

For some years Liquids Research Ltd. has produced magnetic nanoparticles in water for applications in magnetic hyperthermia with the trademark HyperMAG®. These particles are produced via the same technique described in Section 6.1 but following washing, they are subsequently treated with a solution containing an Fe^{2+} salt. The salt is added to the particles in an alkali solution in a dilute concentration with moderate heating. The HyperMAG® particles are then produced via this technique and from the data shown in Section 6.5 and in particular the data relating to the magnetic hyperthermia effect in both liquid and solid forms, do exhibit significant hysteresis heating. This arises because the addition of further Fe^{2+} ions restores the ratio of the $Fe^{3+}:Fe^{2+}$ resulting in a significant increase in the resulting saturation magnetisation. The particles are also significantly larger than the standard ferrofluid particles which can be seen in Fig. 6.17. This leads to these particles having a significantly higher energy barrier (KV) to reversal than would be the case for the standard 10 nm materials. Because of the volume dependence of the energy barrier to reversal, a particle of diameter 15 nm has an energy barrier approximately 3.4 times higher than would be the case for a particle with a diameter of 10 nm. HyperMAG® particles have been shown to reduce the size of tumours in mice at 435 kHz (Kossatz *et al.*, 2014).

As can be seen in Fig. 6.17 these particles also tend to have a significant degree of elongation as well as a higher value of M_s. Because of the significant elongation where it appears that some particles have fused together and the dependence of the shape anisotropy constant on the square of the magnetisation, this also creates a significant enhancement to the anisotropy constant K and consequently the energy barrier to reversal increases significantly. This results in a significant enhancement to the hysteresis heating and because these particles are subsequently embedded in polymer spheres, all stirring effects and aggregation of the primary particles themselves is prevented. The presence of chemical radicals on the surface of the polystyrene spheres also facilitates further functionalisation so that the spheres can be designed to be attracted to certain chemical species within the human body.

Figure 6.24(a) shows magnetic hysteresis loops measured with a BH loop tracer at a frequency of 111 kHz and a maximum applied field of 180 Oe for a sample of HyperMAG® C nanoparticles encapsulated in polymer spheres. Figure 6.24(b) shows the SAR data at different frequencies for that sample as well as for a HyperMAG® C sample in a liquid environment, i.e. without encapsulation. The data shows that at low fields the HyperMAG® particles in the liquid state produce a higher level of heating via the area of the hysteresis loop and stirring. However when the particles are embedded in polymer spheres within a liquid environment i.e. water, and exposed to a marginally larger field the generation of heat by magnetic hysteresis alone exceeds that produced by the particles in the liquid state.

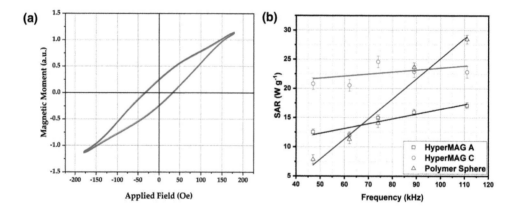

Fig. 6.24: (a) Hysteresis loops measured at 111 kHz and 180 Oe for a sample of HyperMAG® nanoparticles encapsulated in polymer spheres and (b) SAR data for that sample as well as for the same nanoparticles in a liquid environment at a range of frequencies.

There is a further issue with all magnetic hyperthermia treatments in that the application of a large ac magnetic field to the human body is known to cause pain due to the induced eddy currents in conducting tissues. The pain threshold (PT) is defined as the product of the field amplitude H_a and the frequency of operation f, i.e. PT = $H_a f$.

In the commercial system NanoActivator® produced by the company Mag Force AG, the peak field available is 225 Oe (18 kA/m) at a frequency of 100 kHz (Gneveckow *et al.*, 2004). This field would be more than adequate to close the loop for the polymer spheres containing HyperMAG® particles shown in Fig. 6.24(a). However it is reported that for patients receiving cranial therapy 170 Oe at 100 kHz was the tolerable limit (Maier-Hauff *et al.*, 2007). For the groin it was only 62.5 Oe (5 kA/m) (Johannsen *et al.*, 2007).

With hysteresic heating it is possible to operate at a reduced frequency so that the field amplitude can be increased. However reducing the frequency will also reduce the coercivity as discussed in Section 3.5. Hence adjustments to the particle size and shape may be required. Additionally longer treatment times may be necessary as heat loss via the bloodstream will be greater.

We believe that this modified version of an entity such as a polymer sphere embedded within which are particles having significantly improved properties to generate magnetic hysteresis, may allow for a dosimetric strategy for magnetic hyperthermia to be defined. Of course the behaviour of such polymer spheres would in no way be dependent on the viscosity and other aspects of the environment to be found during an in-vivo therapy. This work providing details of the measurements and the production process for the particles which are now produced by Liquids Research Ltd. were published during the production of this text (Vallejo-Fernandez *et al.*, 2023). The concept of producing a hysteresis loss only material based on these ideas have resulted in the application for a UK patent (O'Grady, 2022).

6.6 Bio-Medical Applications

6.6.1 Separation and Localisation

Over the years there have been a number of proposed and to some extent implementation of magnetic nanoparticles for applications such as drug delivery in-vivo and cell separation in-vitro. In both these cases it is the force that can be induced on a magnetic nanoparticle either attached to a drug or a cell, with the field gradient from a magnet used to localise an active pharmaceutical or separate out cells or proteins, that is possible. However such applications rely only on the magnetic moment of the particle that gives rise to the induced force hence the intrinsic complex properties of the particle play little or no part in the application. Based on this requirement ideally one would like to use material having a high magnetisation and that would involve the use of metallic ferromagnets rather than the common ferrimagnets. However because metallic nanoparticles are known to be highly catalytic and in the case of nickel very toxic, it remains the case that most magnetic nanoparticles for use in these areas are composed of magnetite.

Of course to get the nanoparticles to attach to the proteins, cells or other entities it is necessary that they are functionalised with a number of generally organic molecules which then form the attachment for the entity to be separated or delivered. There are a number of companies worldwide that produce magnetic nanoparticles with specific binding agents for these purposes. Examples of such companies producing these materials are Merck KGaA and 5M BIOMED.

One of the most commonly used magnetic nanoparticle systems for applications of this type are magnetite particles coated with the simple sugar dextran. This sugar comes in a number of molecular weights and the particles are produced with the sugar coating so that specific binding agents can be attached to target certain biological entities (Molday and Mackenzie, 1982). One application of particular interest is the use of dextran coated magnetite to attach to lysosomes from within cells which can then be extracted. This application of magnetic nanoparticles has proven to be particularly effective (Walker and Lloyd-Evans, 2015).

In 2003 the UK Institute Physics Journal of Physics D published three related and consecutive review articles detailing the known and potential applications of magnetic nanoparticles in biomedicine (Pankhurst *et al.*, Tartaj *et al.*, Berry and Curtis, all 2003). These three linked reviews detail the physics of the range of applications, chemical synthesis methods for their preparation and the functionalisation of such particles for the requirements of biomedicine. The interested reader is referred to these articles and a subsequent set of updated reviews by the same authors published in 2009 (Pankhurst *et al.*, Roca *et al.*, Berry, all 2009).

6.6.2   MRI Contrast Enhancement

MRI contrast enhancement is one of the most commonly known and widely used imaging techniques for the inside of the human body. The technique relies on the proton relaxation principally within water molecules, that occurs if the system is exposed to a very large DC field typically of the order of 2 T which is subsequently excited by a radio frequency pulse at a much higher frequency close to the Larmor precession frequency of the protons which occur at 42.57 MHz. This gives rise to two relaxation times commonly denoted T_1 and T_2. Both relaxation times are shortened if there is a material with a high magnetic moment in the region to be imaged. This is most commonly achieved by the use of paramagnetic gadolinium iron complexes which can be injected intravenously into the site where enhanced imaging is required. However gadolinium as a heavy metal is somewhat toxic and can only be used in relatively small quantities and infrequently. The alternative is to use superparamagnetic nanoparticles which if inserted into the site where imaging is required, produces a substantially locally perturbing dipolar field which primarily affects the relaxation time T_2^* according to

$$\frac{1}{T_2^*} = \frac{1}{T_2} + \gamma \frac{\Delta B_0}{2} \tag{6.17}$$

where T_2^* is the modified value of the normal relaxation time T_2, γ is the gyromagnetic ratio, $\gamma = 2.67 \times 10^8 \; rad \cdot f^{-1} \cdot T^{-1}$. ΔB_0 is then the change in the field locally due to the dipole moment on the particles.

Other than the gadolinium salts it is possible to alter the value of T_2 using sugar dextran coated iron oxide particles due to their bio-compatibility with the body which are subsequently discharged via the liver.

The case of such particles were quite common until approximately the year 2000 when there were reports of the particles that were up to 30 nm in diameter, inducing thromboses in the lower limbs. Subsequently their use in MRI contrast enhancement has reduced to avoid this effect but they are still commercially available. The use of nanoparticles for MRI contrast enhancement is discussed in detail in the reviews of Pankhurst *et al.* (2003 and 2009).

6.6.3   Magnetic Imaging

In a relatively recent development magnetic nanoparticles have been used for tracking the metastasis (spread) particularly in the case of breast cancer. The nanoparticles are injected near the site of the tumour in very small quantities and a device based on the measurement of magnetic susceptibility is used to track the spread of the nanoparticles through the tissue into the lymphatic system. Such a system is obviously preferable to using tracking based on radioactive substances and at present is in its infancy but is likely to develop in future. The particles used for this application is the former MRI contrast material called Resovist which is a form of dextran coated magnetite discussed in Section 6.6.1. The material achieved approval from the US Food and Drug Administration and other health agencies some years ago.

During the preparation of this book the UK based company Endomagnetics Ltd. has obtained approval from the UK regulatory authorities for the use of this technique in our National Health Service and has been approved by the US FDA (www.endo magnetics.com).

6.6.4   Magnetic Polymer Microspheres

It is little known in the magnetic community but polymer microspheres containing magnetic nanoparticles are very widely used for the separation or concentration of various entities such as proteins, viruses, etc. in vitro. The world market for these materials is astonishingly large totalling something of the order of \$1.2B in 2019. Again there is no specific requirement for magnetic properties other than the particles being superparamagnetic. Generally magnetic nanoparticles prepared via processes similar to that described in Section 6.1 are used. However in certain circumstances additional growth processes may be used and to increase the individual moment on a given nanoparticle and to improve their dispersibility. The starting point for the preparation of such materials again uses oleic acid coated magnetite and the polymer material is generally some form of polystyrene.

Hence particles prepared as described in Section 6.1 are transferred into a carrier such as styrene containing an inhibitor to prevent polymerisation at an unfavoured point in the process. This transfer usually involves the use of a second surfactant because oleic acid is incompatible with styrene. Once a stable ferrofluid has been generated in styrene the material is then emulsified in water using usually a high shear rotor stator mixer which will produce droplets within the water ranging in size from 0.2 μm up to about 5 μm. An initiator is then added to the mixture which starts the polymerisation process which requires heating under various regimes depending on the exact composition, and can take as long as 24 hours. The resulting polymer microspheres have carboxyl $(COOH)^-$ on their surface due to the excess oleic acid in the original ferrofluid. The presence of these active radicals enables other radicals such as amine or hydroxyl groups to be attached via a chemical process. The resulting polymer spheres are thoroughly washed and often filtered to remove any dust or larger polymer entities. Generally the polymer spheres contain about 40% by weight of magnetite such that they can be separated from water with a permanent magnet in a few minutes. An SEM image of such spheres produced by Liquids Research Ltd. is shown in Fig 6.25.

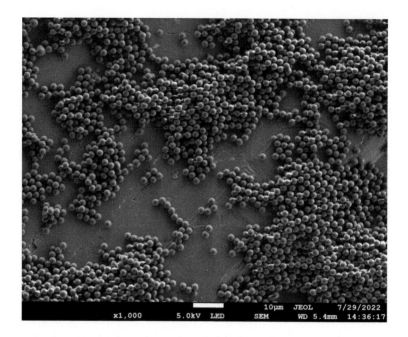

Fig. 6.25: SEM image of polymer microspheres produced by Liquids Research Ltd.

The size distribution of the polymer spheres is generally quite broad depending on the emulsification process but none-the-less it is simply a requirement that the polymer spheres are functionalised to the correct degree so as to attach to the entities that are required to be separated. The separation process generally involves flowing the resulting solution containing the magnetic polymer spheres over a relatively strong permanent magnet which is subsequently removed from the side of the vessel or channel used, thereby allowing the recovery of the materials.

6.7   Magneto-Rheological Fluids

There is a further form of magnetic liquid which is coming into increasingly common use. It is not a true ferrofluid but rather a liquid which has been engineered so as to show a dramatic change in viscosity when exposed to a magnetic field. These liquids, known as Magneto-Rheological Fluids (M-RFs) find quite extensive applications at the time of writing in road wheel dampers on high end luxury cars but increasingly their use is migrating down from cars such as the Range Rover Evoque and Rolls Royce into expensive family cars such as high-end BMW and Audi vehicles. They also find application in the engine mountings of certain generally expensive cars where their use with a feedback system allows for noise suppression and vibration suppression generally from an internal combustion engine.

Interestingly the first widespread application of dampers based on M-RF was as a mounting under the seat in heavy goods vehicles and other large vehicles such as buses. This is the case because drivers spending many hours in such a vehicle were found to have a non-physical form of lower back pain which eventually was identified as coming from the constant vibration of the seat which was damaging nerve ends in the spine in a similar way to high intensity hydraulic tools causing what was known as white finger in the mining industry. The use of an M-RF removes the excessive vibrations.

M-RFs are produced using magnetic particles known as carbonyl iron particles or CIP. These are particles with a size ranging from 1 μm to 10 μm which are produced by the decomposition of iron pentacarbonyl ($Fe(CO)_5$). Iron carbonyl exists as a liquid which is relatively volatile and it is decomposed in a vapour phase with the largest producer being the BASF company of Ludwigshafen in Germany, who offer a wide range of grades of this material. The particles are in some way doped with other elements like molybdenum because at this size they are multi domain particles and if the damper is to relax quickly when the level of vibration varies, it is necessary to ensure that the particles are magnetically soft, essentially having zero remanence and coercivity. The range of additives used can be found on the BASF website. An SEM image of CIP particles is shown in Fig. 6.26 (https://aerospace.basf.com/carbonyl-iron-powder.html).

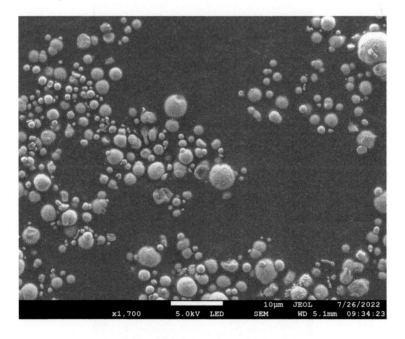

Fig. 6.26: SEM image of iron carbonyl particles used in the preparation of M-R fluids.

To form the M-RF the CIP particles are mixed under high shear with a number of additives included in the oil which is usually a relatively inexpensive polyalphaolefin (POA). The additives used include a dispersant to separate the particles in a similar manner to that described for a ferrofluid, but because the particles are predominantly iron, a number of other additives are included such as antioxidants and antiwear agents which are similar to those used in normal lubricating oils. The components are mixed under very high shear using a saw-tooth blade. Hence the vessel containing the mixture has to be clamped firmly in place. In addition to the CIP particles there are also added a number of micron sized clay particles. These are insoluble in the oil but when a field is applied and the CIP particles attempt to clump together the clay particles get trapped between the CIP particles forming a rigid structure within the liquid. Once the field is turned off, the CIP particles readily separate and hence the mixture is to some degree stable against the effects of the magnetic field.

Of course given that the density of iron is around 9 g/cc and the density of the remaining components is less than 2 g/cc, some sedimentation inevitably occurs in the absence of a magnetic field. However the degree of sedimentation is less than 20% over a period of a year and importantly, when the fluid is formulated correctly, the sediment that forms is a soft sediment that readily disperses once the liquid is sheared again. The structure of an M-R fluid when the field is applied is shown schematically in Fig. 6.27.

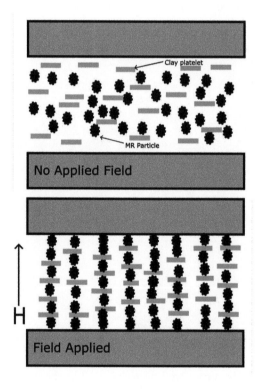

Fig. 6.27: Schematic of the structure of an M-R fluid with (bottom) and without (top) a magnetic field applied.

7
Magnetic Recording

The first report of a magnetic recording system was made as long ago as 1899. The device was a wire recording system invented by the Danish engineer Valdemar Poulsen who recorded information by wrapping a coil around a length of wire which was then energised with an electrical wave generated from a sound wave with some form of amplifier between the microphone and the coil (Poulsen, 1899). Replay was achieved by simply running the wire through a separate coil which was then fed to some form of loudspeaker. The system was only used for audio recording.

The real development of magnetic recording occurred in the late 1930s and primarily after the end of the Second World War. Recording systems were originally based upon lengths of paper which were coated with some form of slurry of magnetic particles. However the real development of practical magnetic recording systems occurred after the end of the second world war when it became possible to produce relatively long, thin lengths of a plastic tape which could then be coated with dispersions of particles. Almost all such systems were coated with elongated particles of maghemite (γ-Fe$_2$O$_3$) but in the period up to the advent of video recording and subsequently digital storage on tape, similar elongated particles were invariably used.

Whilst such systems are relatively little used in domestic settings, at the time of writing they are still in use as archival storage of large data sets. It is worth briefly looking at the history of the development of what remains in certain aspects the highest technology in use at this time as it often illustrates the basic physics of the process.

7.1 Tape Recording

7.1.1 Analogue Recording

Ignoring the very early attempts at tape recording, the first practical audio recording systems were based upon a coating of elongated γ-Fe$_2$O$_3$ particles which were dispersed in solvents with a number of polymeric resin systems which could be coated at relatively low thicknesses. The early systems recorded a single track of an audio signal onto the magnetic tape, but rapidly evolved into multitrack systems using an array of recording heads. As the magnetisation was required to be stored in the plane of the tape to avoid demagnetising effects, it was necessary to use recording heads in the form of a horseshoe with a coil of wire wrapped around the partial loop. Figure 7.1 shows a schematic of such a horseshoe shaped head.

Of course the use of a horseshoe shaped head such as that shown in Fig. 7.1 generates only a moderate fringing field outside of the gap and assuming that the gap length is $2g$ and the flux path length in the core is l_c then the efficiency of the head is

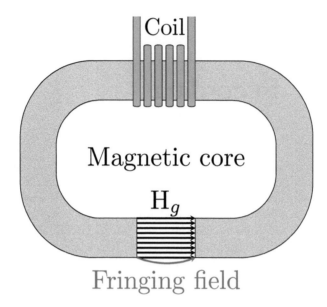

Fig. 7.1: An early horseshoe head for in-plane recording with a small fringing field for writing.

the fraction of the magnetomotive force ($mmf = ni$) which generates the field in the gap. The head efficiency η, is the fraction of the magnetomotive force that appears in the gap and is given by

$$\eta = \frac{2gH_g}{2gH_g + l_cH_c} \tag{7.1}$$

where H_c is the field in the core of the material and H_g is the field in the gap. The field at the surface of the tape is of course not the same as the field in the gap as it is only the fringing field that is available to magnetise the tape. For this reason the tape head of necessity has to run in contact with the tape and in early systems a felt pad was used to press the tape against the head. This resulted in significant wear on the tape and to some extent on the head. Hence the binders used on the tape had to be a combination of what are known as soft and hard polymeric resins to provide the tape with the required durability. The soft resins were generally some type of vinyl mixed with a hard polyurethane resin to cope with the level of wear.

A second significant disadvantage occurred in the replay mode when the same head was in general used but again the head was only able to detect the fringing field given off by the magnetised regions of the tape.

A further disadvantage in recording a signal representing sound on a material magnetised in plane goes back to the nature of magnetisation reversal in magnetic particles. Reference to Fig. 7.2 shows the normal magnetising curve for a particulate or indeed a domain based magnetic material which is highly non-linear. Given this non-linearity, the volume of the sound played back would not represent accurately the

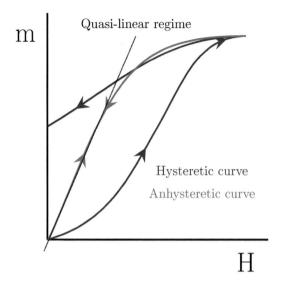

Fig. 7.2: The DC hysteretic magnetising and AC anhysteretic magnetising curves showing the transition from a non- to quasi-linear low-field response.

volume of the signal recorded. This problem was overcome by simultaneously applying a high frequency AC signal to the head along with the quasi-DC signal generated by the sound. Again as shown Fig. 7.2, this results in a significantly linear region which is how the sound was recorded. This same anhysteretic recording technique was in use for all analogue recording including that of video recording, until the invention of digital recording systems.

The particles used in all tape recording systems until the late 1970s were based on elongated particles of originally maghemite (γ-Fe_2O_3) and from about the late 1990s, metal particles. These type of particles were ideal for magnetic recording as they were relatively inexpensive and easy to produce and had predominant shape anisotropy such that the coercivity could to some degree, be controlled. The synthesis for all the elongated iron oxide and metal particles were based on the pseudo-morphic reduction of the mineral goethite (FeOOH). This mineral can be produced via treatment of a solution of ferrous sulphate ($FeSO_4$) with sodium hydroxide (NaOH) and the bubbling of air through the solution and heating. Goethite grows naturally in the form of elongated needles. The goethite needles are then heated to relatively high temperatures often in the presence of other metal salts, which prevents sintering during the reduction of the goethite to hematite (α-Fe_2O_3). Hematite is antiferromagnetic and there is no direct transformation that leads to the magnetic phase of this oxide. The process to form maghemite is by heating the hematite particles in a hydrogen atmosphere which reduces the oxide to the magnetic iron oxide magnetite (Fe_3O_4) which when re-oxidised through to Fe_2O_3, results in the formation of the γ phase known as maghemite which is ferrimagnetic. It was the elongated particles of maghemite that were used for all audio recording but had a limit to the coercivity independent of the particle elongation, in the range 280-320 Oe. The reason why the coercivity was limited and largely independent of the particle elongation, derives from the reversal occurring incoherently via the fanning mechanism discussed in Section 2.5 (de Witte *et al.*, 1990). A coercivity of ~300 Oe was sufficient for all audio recording.

With the invention of the video recording system a much higher coercivity was required so as to store the much higher frequency signal used in television broadcasting. Initially the only material that could generate the required coercivity typically in the range 600-800 Oe, were elongated needles of the ferromagnetic chromium dioxide (CrO_2). This oxide does not derive from goethite but via a fairly complex chemical process and again naturally grows in the form of needles. However CrO_2 has the disadvantage of being abrasive and also under conditions of relatively high humidity, can result in the formation of the highly corrosive chromic acid. This had serious implications for the recording head particularly if a freeze frame was used for any length of time.

A much better solution was discovered in 1974 by Umeki *et al.* of the TDK company in Japan who took the original γ-Fe_2O_3 particles and coated the surface with a layer of cobalt ferrite. Tapes made from this material induced head wear that was only 20% of that caused by CrO_2 whose use was largely discontinued (Umeki *et al.*, 1974).

This resulted in particles that had a coercivity similar to if not greater than that of CrO_2. At the time there was great controversy as to why the anisotropy and hence the coercivity of the particles was increased to such an extent by a layer of cobalt ferrite that was only a few nanometres thick. It was eventually resolved that the roll of the cobalt ferrite was to smooth the particle surface.

Particles of γ-Fe_2O_3 naturally grow with a porous surface resulting from the dehydration of the original goethite. Such pores and perhaps asperities on the particle surface result in the nucleation of magnetisation reversal due to the strong demagnetising field occurring at such sites. Because of the surface energy of the particles the cobalt ferrite would naturally grow at such sites thereby preventing the nucleation of magnetisation reversal. This discovery also solved the mystery as to why the coercivity of the original γ-Fe_2O_3 particles did not depend significantly on the degree of elongation of the needles and hence their shape anisotropy. Using surface doped γ-Fe_2O_3 coercivities of up to 800 Oe were achieved.

With the advent of recording at yet higher frequencies particularly in hand held camcorders and for professional use, a higher coercivity was required than that which could be achieved using any oxide. Hence to generate a higher coercivity it was realised that there would be a necessity to use particles based on a metal due to the shape anisotropy being dominated by the saturation magnetisation of the material as shown in eqn 7.2.

$$K_s = \frac{1}{2}\left(N_a - N_c\right)M_s^2 \tag{7.2}$$

where K_s denotes that the anisotropy is due to the shape and N_a and N_c are the demagnetising factors along the short and long axis of the particle respectively. However the value of the shape anisotropy is dominated by the saturation magnetisation since it is raised to the power 2.

The transition from using particles based on oxides to the use of metal particles presented significant challenges for the industry. The method of preparation of the metal particles proceeds in a very similar way to that used for γ-Fe$_2$O$_3$ but with a number of key additional steps. Different manufacturers used a range of temperatures for the initial dehydration of goethite but also it was necessary to use different and more elaborate coatings on the surface of the goethite. This was necessary so that once the particles were reduced to elemental iron, they could be selectively re-oxidised to put a coating on the surface of the metal particles which then inhibited subsequent oxidation of the core of the particle which consisted of elemental iron perhaps with some addition of cobalt.

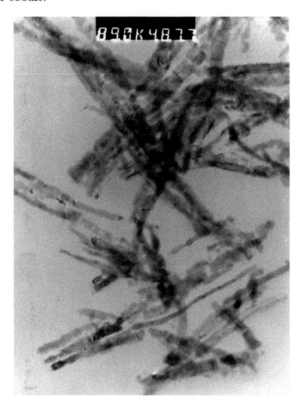

Fig. 7.3: Micrograph of typical metal magnetic tape particles.

The full commercialisation of metal particle tape was dominated by one manufacturer, Dowa Electronics of Japan. Their particles had a surface layer consisting of a mixture of aluminium and yttrium which when oxidised, produced a coating that was relatively impermeable to the oxidising effects of air and in particular moisture. These unusual particles were also much more readily dispersed into the binders than was the case with materials that were produced by other manufacturers and hence became the industry standard for devices such as hand-held camcorders and eventually digital recording standards for high density storage of data. An image of particles produced by Dowa is shown in Fig. 7.3 and as can be seen the particles are very porous and are almost like an agglomeration of much smaller entities. The reason why such a structure produced greater oxidation resistance is not well understood but clearly the surface coating was responsible for these effects and also presumably for their ease of dispersion. Metal particles produced in this way have coercivities that can be as high as 3 kOe and came to dominate the market (Hisano and Saito, 1998).

7.1.2 Digital Recording

The need for higher coercivity in order to increase the recordability of high frequency signals is to some extent obvious because a higher frequency signal will result in regions magnetised in opposite directions being closer together spatially on the surface of the tape. This fact becomes much clearer when one considers the transition to digital recording which began in the 1990s. Figure 7.4 shows clearly the need for higher coercivity in order to store digital data as high density.

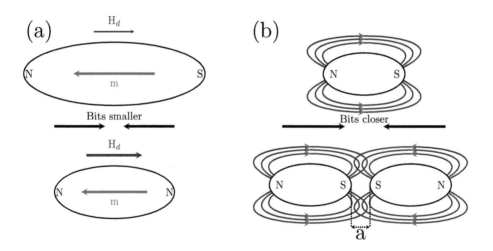

Fig. 7.4: The increased demagnetising field H_d as bits are (a) made smaller and (b) brought closer together.

In order to achieve high density storage in a digital format clearly the bit length needs to be made shorter. This increases the demagnetising field (H_d) from each magnetised region acting upon itself. However there is also a need to bring the bits and indeed the tracks on the tape closer together which again increases bit-to-bit demagnetising effects on all the stored data. Therefore a higher coercivity is required to resist the demagnetising fields. It should be noted that in addition to the demagnetising field its presence can also lead to long term loss of data due to time dependence effects discussed in Section 3.5 as the energy barrier to reversal is reduced by the factor $(1 - H_d/H_K)^2$.

There is another unusual phenomena that can occur in tape recording due to the tape being coiled around a spool. Under these circumstances and particularly prevalent in analogue recording, it is possible that the stray field from one layer of tape in the spool magnetises an adjacent layer. In particular when playing music if there was for example a clash of cymbals in the recording, then listening carefully to the tape would produce an echo of this high volume and hence high magnetisation event, a few seconds later due to *print through*, of the signal from one layer to another. This phenomenon is less common in digital recording but does show that there is a significant time dependence effect due to stray fields in all tape recording systems. Reference to Fig. 7.4(b) shows that in order to increase data density and hence data capacity, requires not only the ability to write shorter bits, but also to be able to switch the direction of magnetisation between neighbouring bits. This separation labelled with the parameter "*a*" is called the transition width and is related to the width of the switching field distribution and to a write head parameter as shown in eqn 7.3.

$$a = \frac{M_r\delta}{H_c}\left(d\left(1 - \frac{S^*}{\pi Q}\right) + \frac{d\left(1 - S^*\right)}{\pi Q} + \frac{M_r\delta}{H_c}\right) \tag{7.3}$$

In this expression the first term relates to the remanent magnetisation and the coating thickness δ, $M_r\delta$, divided by the coercivity. The terms in the brackets relate to the switching field distribution, $1\text{-}S^*$ (see Section 3.6) and the parameter πQ which is a parameter defining the head field. The parameter d is the head to medium separation. Whilst obviously a narrow switching field distribution is desirable but difficult to control, most of the physics that determines the transition width is given by the remanent thickness product divided by the coercivity.

A higher coercivity allows the transition width to be reduced but also the implications of eqn 7.3 is that a material with low remanence is required because it is the remanent magnetisation in the bit that determines the demagnetising field and additionally that is related to the thickness of the recording layer. Of course reducing the remanence of the media also reduces the output signal which has to be read. Hence there is a clear and delicate compromise to be sought between the bit length and the magnetisation of the recording medium itself. Equation 7.3 is based on the first realistic model of the write process in magnetic recording originally due to Williams and Comstock (1971).

The ability to increase the density in digital recording depends on being able to fabricate a suitable write head with a narrow gap giving the largest field possible but also on the sensitivity of the read head device. Furthermore the ability to control the track width again relies on the ability to fabricate a suitable head. The issue with the write field was significantly improved not only by narrowing the gap in the head but also by depositing a strong diamagnetic material such as bismuth into the gap thereby forcing the flux out of the gap towards the surface of the tape.

The next significant development which was applied to both tape and disk systems relied on the work of David Thompson and his group at IBM who first introduced the use of a thin film magnetoresistive head (Thompson *et al.*, 1975). These heads were based on the well known anisotropic resistance effect discussed in Section 5.1 and whilst the change in resistance was relatively small at around 2%, the ability to fabricate a smaller device allowed for a significant advance in increasing data density. For the case of a magnetoresistive sensor, the thin film was placed on edge above the tape and hence the effective width of the read head now becomes related to the thickness of the magnetoresistive layer deposited on a substrate. Hence a much narrower read head with far higher resolution was possible with this technology.

The change in resistance ($\Delta\rho$) varies according to the direction of the field which gives rise to the effect in accordance with

$$\frac{\Delta\rho}{\rho} = \frac{\Delta\rho}{\rho_{max}} \left[\frac{2}{3} - \left(\frac{H_y}{H_a} \right)^2 \right] \tag{7.4}$$

where ρ is the resistivity and H_y and H_x are the fields from the bit along and across the medium.

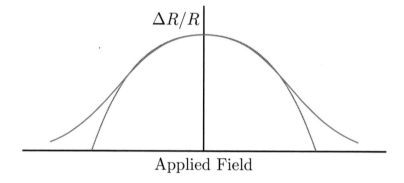

Fig. 7.5: The response of an MR-head to an applied field (red line) compared to an ideal quadratic (blue line).

For a head with transverse anisotropy and a field directed in the y, i.e. in-plane direction, this implies that $H_y/H_{ha} = sin^2(\theta)$. A quadratic response of this form is non-ideal for a sensor to determine the transitions at the end of the bits because by definition it is non-linear. Figure 7.5 shows the variation of the magneto-resistance effect with the magnitude of field emanating from the recording medium perpendicular to the plane of the medium. As is clear from Fig. 7.5, in order to linearise the response requires the thin film sensor to have a bias field applied to it which was achieved by depositing thin film magnets to bring the sensor into its quasi-linear response region. Furthermore because of the improved sensitivity of this read head, it was also found that the head was able to detect not only the bits it was reading, but also be affected by the stray field from neighbouring bits. This difficulty was overcome by incorporating shield layers in the head made from a soft ferromagnetic material so that the flux from neighbouring bits would not interfere with the magnetoresistive sensor which was generally produced using a thin film of permalloy ($Ni_{80}Fe_{20}$) approximately 100 nm thick. This development is important for modern hard disk drives because similar techniques are used in the design of modern hard drive read heads based on the GMR or TMR effect discussed in Sections 5.3 and 5.4. The overall design of the structures used are shown schematically in Fig. 7.6.

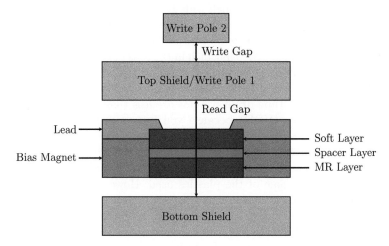

Fig. 7.6: A schematic of a read/write head for tape applications, showing the bias hard magnet and shielding. Not that one pole of the write head forms the shield for the read head.

Of course modern tape drive systems are still limited by the parameter $M_r\delta/H_c$ and it was found that the coercivity of the metal particles described above was limited to values generally below 3 kOe. As will be discussed in detail in Section 7.3 it is now well established that recording information in the plane of the film is inherently limited by the demagnetising field from the bits and from neighbouring bits and as is the case in hard disk recording, a transition was made in tape media to recording data perpendicular to the plane of the medium. This could not be achieved with elongated particles because of their tendency to chain together producing a very rough surface.

The obvious candidate to replace elongated particles is barium ferrite where the material grows naturally with a hexagonal structure with very strong magnetocrystalline anisotropy along the c-axis. Unfortunately barium ferrite grown directly by a precipitation process grows in the form of hexagonal platelets and these platelets, because of the flat surfaces, have a tendency to stack rather like a pile of dinner plates. Again this will produce a rough surface making them unsuitable for use.

However techniques have now been developed to grow barium ferrite by first growing spherical particles of magnetite or maghemite and subsequently depositing a layer of barium oxide (BaO) on the surface of the particle. Careful annealing and with the use of anti-sintering agents, allows the double layered particle to become transformed into barium ferrite but retaining the spherical shape which reduces or completely removes the probability of stacking (Jalli *et al.*, 2009). Hence the highest density tape formats available in the world today operate in perpendicular mode with GMR or TMR read heads similar to those that will be described for the state-of-the-art in hard drives, and operate from a single spool cassette which is loaded into a tape reader where a second spool is used to drive the tape to be read.

The most common format for tape drive systems today is known as LTO9 which refers to linear recording along a one inch wide tape format. The tape is controlled by an elaborate servo track written down the middle enabling a very high track density and with the high coercivity of up to 4.5 kOe of the barium ferrite particles allows a capacity of up to 18 Tbyte per cartridge which in the case of barium ferrite, corresponds to a data density of 80 Gbit/in^2. There are even proposals to go beyond the 100 Tbyte cartridge using strontium ferrite again prepared in a spherical form.

It remains the case that tape drive systems are less than 1/5 of the cost/bit compared to a hard disk drive (HDD) because you buy one drive and perhaps many hundreds of tape cartridges. It is also the case that tape drives are approximately 30% less expensive per gigabyte than cloud storage (wwww.techxplore.com).

There do however remain issues with the use of tapes particularly with regard to the long term stability of some of the plastics used. However large data warehouses are now constructed with controlled atmospheric conditions with very low humidity and perhaps even reduced temperature and certainly with control over the level of light and in particular UV light, which can cause the plastics to degrade. Under those conditions the data storage capability of a tape in terms of life is similar to that of a hard disk system.

The one issue that cannot be overcome is that of access time. Clearly to find a set of data on a tape requires the unspooling of the whole tape and re-writing the data onto a hard drive and only then can the hard drive be interrogated at high speed. However there are certain areas where the long term storage of data is required and where access time is not critical. Such areas include the records of parliaments and company records and records of committees and even information such as bank or pension records, where economy is more important than speed. Hence it remains the case that tape recording represents a major industry worldwide, albeit little known.

7.2 Longitudinal Disk Recording

In addition to conventional tape recording described above, in the early days of personal computers there was a need to have a device to hold software and store files due to the lack of capacity of random access memory (RAM) at that time. To this end a floppy disk system was developed which was basically a circular sheet of tape contained within a cardboard or plastic sleeve with a window on it which was either internal or external to early personal computers. Using the disk format had obvious advantages over a tape format in that access times were reduced significantly and the device was capable of storing data at a higher density than conventional tapes available at that time. So called floppy disks again went through a number of generations with increasing coercivity in the particles similar to that used in tape recording. However because of the requirement for ever larger coercivity and the obvious advantages of having a mechanically rigid system, rapidly led to the development of hard disk drives originally based on wet coated γ-Fe$_2$O$_3$ elongated particles but again due to the requirements for increased data density, a transition to a thin film, metal based structure rapidly evolved. Obviously also on a rigid disk it is possible to have a much higher track density and rotation speed.

Once again considering eqn 7.3 which describes the minimum distance between transitions, with a metallic based thin film it was possible to control the remanence of the coating almost as required but importantly the parameter δ could be made much thinner and yet give the same level of output signal as the much thicker wet coated oxide particles. Furthermore by the use of cobalt or more importantly cobalt alloys, it was also possible to create coatings having a much higher coercivity and hence able to support much narrower transitions. Also because the disk is now rigid it is possible to bring both the read and write heads much closer to the surface than was possible in the case of a floppy disk. More importantly it was also possible to rotate a rigid disk at much higher speeds. Irrespective of the type of read head used, the signal induced in the head in the read function depends upon Faraday's Law.

$$\epsilon = -N\frac{d\phi}{dt} \tag{7.5}$$

In eqn 7.5 the induced emf depends upon the rate of change of the flux so that if a disk written with an array of bits is rotating very quickly then automatically $d\phi/dt$ increases due to the shorter time that the bit is exposed to the read device. Initially hard disk drives rotated at 3600 rpm which was tied to the mains frequency of 60 Hz used in the United States. However in subsequent generations of hard drives the rotation speed was increased eventually to be as high as 15,000 rpm but more commonly 7550 rpm. Figure 7.7(a) shows the schematic diagram of a longitudinal recording system and 7.7(b) shows a photograph of a hard disk drive originally developed for use in a digital audio recording device used to store and play music.

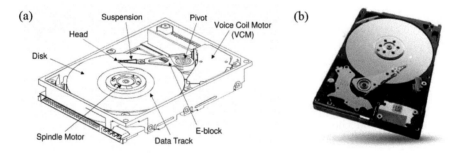

Fig. 7.7: (a) A schematic of a hard-disk drive (Oldham *et al.*, 2008) and (b) an image of a typical Seagate hard-disk drive.

The magnetic materials of choice for early hard drives were cobalt alloys where the coercivity was controlled by using alloys of cobalt (Co) and platinum (Pt). In the case of these alloys the much larger Pt atoms sit in the interstitial spaces within a conventional hcp Co crystal as shown schematically in Fig. 7.8. The Pt atoms located as shown in the figure increased the c/a ratio of the Co crystal thereby enhancing the coercivity and also served to increase the corrosion resistance of the thin film coating. However again being guided by eqn 7.3 there is a requirement to reduce the remanence and hence the saturation magnetisation of the material. It is well known that alloys of Co and chromium (Cr) have a highly controlled reduced value of saturation magnetisation. In fact an alloy containing approximately 40% Cr becomes paramagnetic at that concentration. Hence the use of Cr in the alloy became the norm.

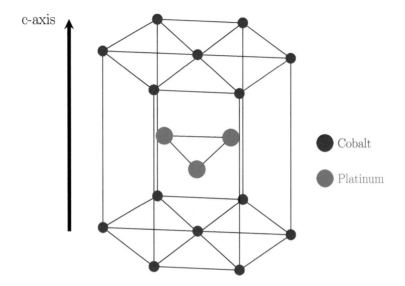

Fig. 7.8: The hexagonal crystal of CoPt where the large Pt atoms (red) cause a stretch of the c-axis and increase the crystal anisotropy.

The typical substrates used prior to the invention of laptop computers was an alloy of aluminium and magnesium (Al/Mg) which was produced from sheets of the alloy which were then stamped to produce disks. To ensure a perfectly flat surface so that the head could fly in close proximity, it was necessary to add a smoothing layer which was generally an electrolessly deposited layer of nickel phosphorous (NiP). There was also the requirement to somehow align the c-axis of the Co-alloy grains circumferentially on the surface of the disk so as to obtain magnetic behaviour as close as possible to the ideal aligned case given by the Stoner-Wohlfarth model. This was achieved via the rather crude technique of scratching the surface of the disk using a soft pad impregnated with very small particles of diamond typically about 1 micron in diameter which were soaked in water and pressed down onto a rotating disk so as to produce an array of the required circumferential scratches. This technique was partially successful and could result in the production of disks with a squareness of the hysteresis loop (M_r/M_s) of up to 0.85 in the best cases. However this value was in part due to intergranular exchange coupling (see Section 2.6).

However it was also necessary to produce an appropriate seed layer to induce the c-axis of the cobalt grains to grow in plane. After many years of achieving the necessary in-plane growth, the most commonly used seed layers was either Cr or an alloy of this element where the lattice match to the alloy was found to result in what were known as bi-crystal media where the Co (110) plane of the Co-alloy crystal grew epitaxially onto the (200) plane of the Cr underlayer. However this resulted in a 28° out of plane orientation of the Co c-axis. This lattice match is shown in Fig. 7.9.

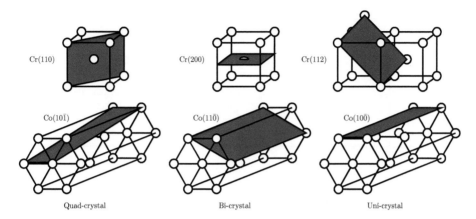

Fig. 7.9: The lattice match of hcp Co to a seed layer.

In a hard disk drive the read and write heads are mounted on a chamfered arm known as a gimbal such that as the disk rotates, the head takes off and flies like an aeroplane above and as close as possible to the surface of the disk. Of course when the disk is not rotating the head has to land somewhere and this can result in mechanical damage of the disk. Over the development of longitudinal recording a number of surface coatings were used to try and protect the disk from damage by the head structure and eventually technology was developed such that it was possible to deposit an ultrathin layer of carbon which grew with a diamond like structure (DLC) making it extremely hard and durable. This was achieved by the simple expedient of decomposing acetylene (C_2H_2) in a chamber where a plasma was struck which, under the correct conditions, resulted in the deposition of DLC (Weiler *et al.*, 1996). Additionally the surface of the disk was lubricated with a complex near monomolecular layer of a perfluorinated polyether oil (PFPE). These oils are extremely inert and have excellent high temperature performance coupled with the fact that they have very low vapour pressure and hence do not need replenishing during the life of the drive.

Based on these techniques and in particular improvement in growth techniques that enabled the control of the grain size in both disks and heads, almost invariably through the use of seed layers whose grain size could be controlled, resulted in the capacity of hard drives undergoing one of the most remarkable rates of improvement of any technology. Between the introduction of thin film disks around 1975 to the year 2000 the growth rate in the aerial density, i.e. the number of bits usually quoted as being per square inch, increased at a compound growth rate which at times exceeded 60%/year. This remarkable density growth including into the new millennium is shown in Fig. 7.10. Marks on the figure are key technological advancements contributed to and allowed for such a dramatic rate of improvement which at times far exceeded that usually described by the famous Moore's Law used to describe the reduction in feature sizes in silicon chips.

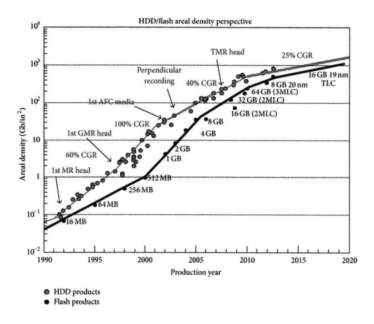

Fig. 7.10: HDD density growth curve between 1990 and 2010 (Marchon *et al.*, 2013).

Due to the advances in growth techniques and automation of the production process, this was also accompanied by a dramatic cost reduction in terms of the cost per gigabyte of data that was available not only for industrial and commercial use but also on sale to the general public. As can be seen in Fig. 7.10 one of the key advances that drove the growth rate was the development of the spin-valve head which is based on the GMR effect discussed in Section 5.3. As will be discussed in Sections 7.3 and 7.4 current hard drive technology is based on recording the stored data in grains oriented perpendicular to the plane, hence a full description of the read-write head structure will be discussed for the perpendicular case in Section 7.5.

Whilst the development of longitudinal recording resulted in enormous technological advances it also presented a number of significant challenges that did not affect storage on flexible media. As already discussed one of these was achieving the orientation of the easy axis of the single domain particles in a circumferential direction around the disk. At best this problem was never completely overcome so as to give rise to the aligned case in the Stoner-Wohlfarth model. Despite the advances, until the invention of DLC, the issue of head crashes and wear was only fully overcome towards the end of the longitudinal recording era. Critically the use of the fringing field from a right head severely limited the availability of an adequate field to magnetise high coercivity grains thereby imposing an upper limit on the transition width as defined by eqn 7.3.

However the major challenge resulting from the development of thin film rigid disk technologies came about because a thin film consisting of a metal or an alloy is by definition a conducting material. As such the presence of the conduction electrons invariably leads to the existence of RKKY interactions between the ferromagnetic grains which then results in cooperative reversal as discussed in Section 2.6. This can produce a material whose magnetisation reversal is via a quasi domain wall process whereas the requirement for high density storage is for reversal to occur via the rotation of the moments of single domain particles. During this era a great deal of experimental and theoretical work was undertaken with regard to the control of the interactions between the grains in the disk the results of which are still applied to this day in modern hard drives based on perpendicular recording.

On the experimental side it was found that increasing the amount of Cr in the alloy resulted in some degree of grain segregation. This arises because the Cr as an impurity in the alloy grain, tends to segregate towards grain boundaries and as discussed above, this can lead to a paramagnetic or even an antiferromagnetic structure in a few surface atom layers of the grain. This was most beautifully illustrated using high resolution electron dispersive X-ray spectroscopy (EDX) by Futamoto *et al.* in 1998. Using high resolution EDX they produced images which showed the segregation of Cr to the grain boundaries in concentrations of up to almost 30% hence approaching the non-magnetic composition at about 2 nm from the grain surface.

Coincidentally at the same MRS meeting in Boston our group presented data showing the effect of grain segregation on the interactions in thin films via the ΔM measurement discussed in Section 3.8.2 (O'Grady *et al.*, 1998). In our case the grain segregation was achieved by growing ultra-thick (200 nm) Cr underlayers which result in grain growth in a mushroom type structure that then segregates the Co-alloy grains subsequently grown on top of the underlayer. This grain segregation in both sets of results described was accompanied by a significant reduction in the resulting signal/noise ratio emanating from the disk.

The confirmation of the responsibility of cooperative reversal between the grains leading to noise in thin film media was delivered by a number of large-scale computer models that had recently become available in the late 1990s as computer power increased. Notable among these models were those of Zhu and Bertram (1989) who used particles in an hexagonal array and predicted that the RKKY interactions would lead to a quasi-domain structure following a zig-zag pattern. It was shown subsequently that this indication of the nature of the transition was in part an artifact due to the use of the particles on a hexagonal array. More realistic models notably by Chantrell and co-workers (van Kooten *et al.*, 1993) and the group of Victora at the University of Minnesota, where the particles were arranged randomly, also indicated significant transition broadening due to the presence of RKKY interactions (Xue and Victora, 2000).

Complex RKKY interactions can throw up some apparently anomalous results. In a thin film disk where the density of the closely packed grains approaches 90-95% there is also the presence of very strong dipolar interactions between the single domain grains. In principle and as shown by the use of ΔM curves, the exchange interactions dominate the behaviour of the magnetisation reversal process. However in an interesting piece of work done in our laboratories in collaboration with the group of R. W. Chantrell, showed that the presence of grains with a reduced anisotropy, largely arising from crystallographic defects known as stacking faults, can lead to an increase in the localised coercivity. This arises because the localised coercivity over a size range equivalent to a few tens of particles, will seek to minimise the overall energy of the system and the presence of low coercivity or soft grains can result in the formation of micromagnetic configurations of the moments which become more stable than would be the case if all the grains had high anisotropy which to some degree, will resist the reorientation of the moments due to the exchange coupling (O'Grady *et al.*, 1998).

At the time these large-scale computer models were developed it was the first time that such models had produced realistic results that not only agreed with magnetic measurements such as those of ΔM but also gave a reasonable degree of correlation with important application phenomena such as the signal to noise ratio generated by the media and the inability of the media to support an ever increasing on-track density due to the effects of the RKKY interaction. The importance of this result in the development of subsequent media once the transition to storing data perpendicular to the plane of the disk was adopted.

7.3 Transition to Perpendicular Recording

As long ago as 1979 the group of Iwasaki at Tohoku University pointed out the obvious limitations of longitudinal recording and recommended a new paradigm for information storage where the bits of information are stored perpendicular to the plane of the film (Iwasaki *et al.*, 1979). For the magnetisation to be stable in this orientation it is necessary to have a material where the anisotropy field (H_K) is greater than the global demagnetising field which, perpendicular to the plane of a thin film, gives rise to a demagnetising factor (N_d) of 4π in the cgs system. Hence following the Stoner-Wohlfarth model for the perfectly aligned case this gives rise to

$$H_K = \frac{2K}{M_s} \geq 4\pi M_s \tag{7.6}$$

$$K \geq 2\pi M_s^2 \tag{7.7}$$

The criteria arising from eqn 7.7 at first glance seems difficult to achieve. However it had long been known that hcp Co when grown on a neutral substrate tends to develop a natural perpendicular anisotropy due to the higher density of the Co atoms in the basal plane. Co has a relatively high value of saturation magnetisation (1400 emu/cc) and often perpendicular anisotropy does not result. However if the cobalt is alloyed with Cr the saturation magnetisation reduces as discussed in Section 7.2 and it is possible to make CoCr films exhibiting almost perfect perpendicular anisotropy where the Cr level is about 20%. The use of an alloy with a reduced value of M_s also has the advantage of narrowing the transition width as shown in eqn 7.3.

There are obvious advantages to storing data perpendicular to the plane of the disk which arise from the nature of the demagnetising field arising from magnetised regions or bits as shown in Fig. 7.11.

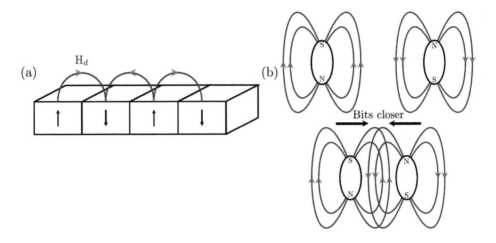

Fig. 7.11: Perpendicular recording bits where the demagnetising field stabilises neighbouring grains.

As can be seen from Fig. 7.11(a) the nature of the demagnetising field with perpendicular orientation reduces the bit-to-bit demagnetising effect but importantly, as the data density increases, the bit-to-bit demagnetising field reduces even further and is often supportive of the required bit pattern as shown in Fig. 7.11(b). For this reason particularly as the coercivity achievable with in-plane recording reached its limit, the transition to perpendicular recording became inevitable and advanced rapidly.

Whilst in principle to store data perpendicular to the plane of the disk would involve a relatively simple transformation of longitudinal recording, in practice it became significantly more difficult due to the need to completely revise the design of the write head. A complete redesign of the disk structure and more importantly a re-engineering of what is known in the industry, as the channel was also required. The channel is the signal processing, error correction and analysis system used to convert what is just a flux pattern into a readable and decodable set of information.

In practice the process of writing data perpendicular to the plane of a disk is relatively simple. Figure 7.12 shows a schematic diagram of how the write process works. A similar write head to the horseshoe used in the longitudinal mode is once again used. However the design of what became known as a single pole head, requires that the write pole be very much narrower than the flux return pole so that the perpendicularly oriented field flux density from the return pole is much lower than that from the narrow single pole head.

The major advantage of a single pole head is now also clear because the growth of an underlayer of a soft material beneath the recording layer now leads to the creation of a magnetic image of the field from the single pole head thereby concentrating the flux over a much narrower region enabling materials with a higher coercivity to be reversed. Again in accordance with eqn 7.3 the fact that materials with a higher coercivity could now be used as the storage medium, automatically narrowing the transition width in

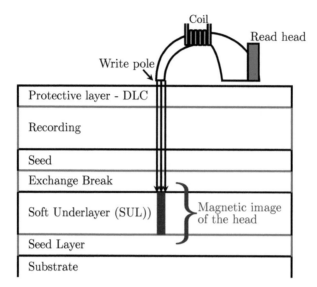

Fig. 7.12: The basic structure of a perpendicular recording medium and the write process, where an image of the write head is focused and produced in the soft underlayer.

the recorded pattern. Whilst this design was in principle easy to implement, in practice it required significant advances in material science both to produce a write head that would work but also to produce a thin film media structure that would have the required properties for high density storage.

There was also an issue in that sputtered films of CoCr are known to have relatively rough surfaces which are far rougher than those when the cobalt alloy grains are grown with the c-axis in plane. There are also further issues whereby the writing of the bits into the recording layer also caused the formation of domains in the soft underlayer (SUL) which resulted in spurious signals when the read head crossed over a domain wall. Hence it took a number of years before all these challenging material science difficulties were fully overcome.

The issue with the read head was in principle much more simple to resolve. Basically in the first instance a GMR but subsequently a TMR read head was used. In the perpendicular orientation the read head has to be reoriented so that the recorded pattern is detected perpendicular to the plane of the thin film sensor. As we will discuss in Section 7.5 this means that the linear bit density along the track is now determined by the overall thickness of the stack from which the read head is composed rather than the limit of lithography which was the case for longitudinal recording. It also became possible to develop and produce integrated read-write head structures which further advanced the data density particularly with the use of focussed ion beam milling (FIB) to shape the single pole head so as to optimise the write field available.

7.4 Disk Structure

7.4.1 Development of Perpendicular Media

In the early stages and the first few generations of HDDs utilising perpendicular record-ing, very similar techniques were used to produce the disk media. This involved the use of similar CoPtCr alloys to form the recording layer. Additional steps were required to induce the required growth of the c-axis of the grains perpendicular to the plane of the film. Clearly different seed layers were required as there was no need to attempt to align the c-axes "in plane". Because hcp Co grows naturally perpendicular to the plane a disordered seed layer was required. Additionally to produce super smooth layers there was no requirement for a smoothing layer on top of the substrate. Hence there was a fairly rapid transition to the use of ceramic glass substrate similar to those used for HDDs for incorporation into laptop computers. The use of seed layers also allowed for extreme control of the grain size. One of the advantages of perpendicular recording is that much higher anisotropies can be used which would allow for the use of smaller grain sizes since the energy barrier to reversal depends upon the product KV.

It is well known that sputtering the element tantalum (Ta) at a high deposition rate results in an almost perfectly amorphous structure due to the very high melting point of this element. Hence it was used to allow other techniques to be used to control the grain size as required. Of course as discussed above, there was a need to use a soft underlayer and a range of alloy compositions based on NiFeX alloys were used by various manufacturers. These were designed so as to limit the degree of domain wall noise that appeared in the media. In many structures a thin antiferromagnetic (AF) layer was grown above the SUL which was found to inhibit domain wall effects in the SUL which would generate spurious signals in the read operation. An intermediate layer typically of ruthenium (Ru) which also exists in an hcp structure and grows with its c-axis perpendicular to the plane, was then grown to form the seed layer for the recording layer. This also served to limit exchange coupling transmitting between grains in the recording layer via the SUL and gave significantly improved perpendicular orientation to the CoPtCr grains in the storage layer. Similar overlayers were used as was the case for longitudinal recording structures i.e. a DLC layer and a lubricant layer of a PFPE.

A number of different structures were used in an attempt to increase the data density further. One of these that was very successful, was the use of a coupled granular continuous (CGC) layer deposited over the top of the recording layer which acted like a keeper on a pair of permanent magnets so as to stabilise the data pattern in the recording layer (Nolan *et al.*, 2011). However the original problem of RKKY coupling between the grains in the storage layer remained an issue as was the case of longitudinal recording. Again varying levels of Cr in the alloy of the storage layer were used in an attempt to inhibit this.

From about 2010 onwards it was established that it was possible to co-sputter the recording layer of the Co-alloy together with silica (SiO_2). As an oxide, silica is an excellent insulator and is also completely insoluble in the Co-alloy. As such the presence of the silica segregated to the edges of each grain. This reduced and when thick enough, completely eliminated the RKKY interaction between the grains. The co-sputtering could be achieved by simultaneously depositing the Co-alloy using DC magnetron sputtering while simultaneously depositing silica onto the surface of the disk but as in insulator this would require an RF sputtering gun.

However many of the sputter deposition targets in use were produced by powder metallurgy where the materials of which the target was composed were produced by first jet milling high purity alloys which were then subsequently pressed together under extreme pressures and heated to produce targets with an existing fine grain structure. These were sufficiently mechanically robust to withstand the effects of the plasma particularly when the targets were water cooled which is normally the case. Hence composite targets containing the Co-alloy and SiO_2 were produced and because of the dominance of the alloy in the target, were still highly conducting and hence would attract the plasma to the surface under the DC bias voltage. However the bombardment of the Ar^+ ions not only sputtered the Co-alloy but simultaneously sputtered the SiO_2 from the compound target.

This technique represented one of several major advances that resulted from the shift to perpendicular recording. Figure 7.13(a) shows a schematic diagram of such a structure where the Ru layer is denoted as the intermediate layer. Because of the almost identical crystal structure and lattice parameter of the intermediate Ru seed layer, the boundary between that layer and the recording layer is quite difficult to see in the cross-section TEM image also shown in Fig. 7.13(c). For clarity purposes this is shown schematically in Fig 7.13(b). The size bar on this cross-section TEM image taken in our own laboratories, also gives an indication of the relative thickness of the various layers. For clarity the recording layer is approximately 10 nm thick.

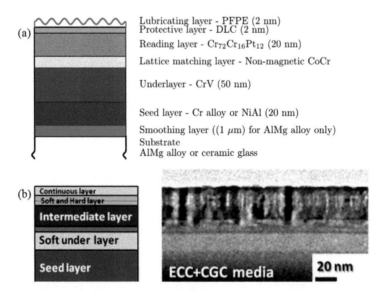

Fig. 7.13: The basic structure of a perpendicular recording media and (b) cross-sectional TEM images of real media (Chureemart *et al.*, 2013).

The development of co-sputtered media and the use of SiO_2 led to rapid and significant advantages in the data density of disks in HDDs. As discussed earlier, one particular major advance was the control or elimination of RKKY interactions thereby narrowing transition widths significantly. However it was found that the incorporation of the SiO_2 also gave an aid to the control of the grain size in the medium and created a significant impact on the coercivity. Figure 7.14 shows a series of hysteresis loops measured in our laboratory. These samples were provided to us via a long-standing collaboration with Tom Nolan of Seagate Media Research in Fremont, CA. Because of the commercial nature of the samples and the requirements of confidentiality, the data shows samples with a high, medium and low level of SiO_2 but the precise levels of silica are not known. Details of the samples and their properties were reported by Chureemart *et al.* (2013).

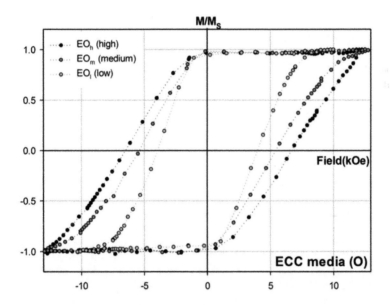

Fig. 7.14: Hysteresis loops for ECC media with varying levels of SiO_2 (Chureemart *et al.*, 2013).

These samples show some quite remarkable properties which are important in the context of understanding the relationship between the deposition method, the magnetic properties and the subsequent potential for use as high density recording media. The first and most obvious observation is that the loop squareness (M_r/M_s) is now very close to 1. This implies that the orientation of the grains is almost perfectly perpendicular and hence the behaviour of these thin films is a very close approximation to the aligned case in the Stoner-Wohlfarth model described in Section 2.4. This therefore is almost the optimum case for the recorded signal. It should be noted here that the data shown in Fig. 7.14 has not been corrected for the sample shape demagnetising effect where the demagnetising factor will be precisely $N_d{=}{-}4\pi$. The data is displayed in this form because this is the effective magnetic behaviour of the disk in the hard drive, rather than the intrinsic behaviour of the material itself.

The second striking feature is that the level of SiO_2 has increased the coercivity significantly from around 4 kOe to more than 6 kOe. There are two possible effects responsible for this significant increase. The first is that using a higher level of SiO_2 will reduce the grain size. However that would be expected to reduce the coercivity whereas it is observed to increase. The other factor which is believed to dominate, is that the presence of the high level of SiO_2 would significantly reduce and possibly, eliminate the RKKY interactions. The presence of RKKY interactions does not intrinsically reduce the coercivity but due to the coupling, the reversal of the grains will be based on a weak link effect whereby the grain with the smallest volume will initiate a reversal and then the RKKY interaction will reverse larger grains that would ordinarily be more stable, in a form of a cascade or cooperative reversal. Hence segregating the grains not only limits the effect of smaller grains with a lower energy barrier to reversal, but also broadens the range of the switching fields as can be seen in Fig. 7.14. Reference back to eqn 7.3 indicates that this broader range of switching fields would broaden the transitions. Hence manufacturers had to seek a compromise between the effect of the RKKY interactions and the desirability for the very high coercivities which were now possible.

Of course for a disk structure having a coercivity of over 6 kOe and a broad switching field distribution such that some grains were switching at fields approaching 10 kOe, could not be switched by the field generated by a single pole head even with the presence of an SUL focussing the field. Hence in this respect the quality of the disk materials available now exceeded the ability to write data. In an ingenious piece of material design, the hard drive industry overcame this challenge by the use of what is known as Exchange Coupled Composite (ECC) media.

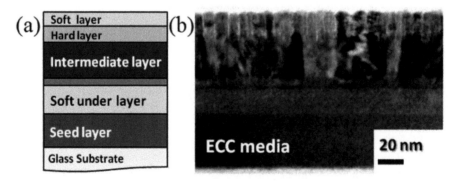

Fig. 7.15: (a) The structure of ECC media (schematic) (b) Cross section TEM of an ECC medium.

In these structures the recording layer shown schematically and in cross section TEM in Figs. 7.15(a) and (b), was divided into two separate layers. The first of these was a high coercivity layer with a value of coercivity of around 5 kOe. A layer with this value of coercivity has a relatively high level of Pt in the alloy thereby increasing the magnetocrystalline anisotropy. As indicated, no head field could be generated to switch this layer. However a second layer of a CoPtCr alloy with a lower coercivity was then grown on top of the high coercivity layer. This layer had a lower Pt content but because the Pt atoms lie in the interstitial spaces within the hcp crystal, there is a perfect lattice match between the softer layer and the hard layer in the basal plane. The write head was now able to switch the softer layer and the direct exchange interaction between the soft and hard layers meant that the soft layer would then switch the hard layer so that the effective write coercivity might be as low as 3 kOe but the storage coercivity was then in excess of 5 kOe. This remarkable piece of magnetic engineering enabled the data density in perpendicular media to increase significantly. At the time of writing (2022) HDDs disks of the ECC media form are still in current production. In Fig. 7.15(b) the space bar indicates the thickness of the various layers. It is not possible to see the boundary between the soft and hard layers in the storage layer due to the precise lattice matching. Again these samples were supplied by Seagate Media Research in Fremont CA.

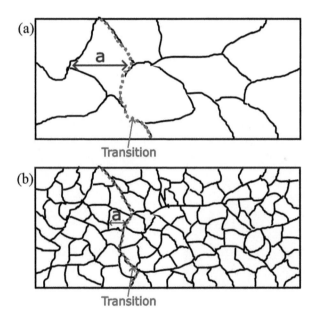

Fig. 7.16: Schematic of the transitions in (a) large and (b) small grained media.

Because of the importance of the use of SiO_2 segregation on the control of grain size and the control or elimination of the RKKY coupling, it is worthwhile looking in more detail at exactly how this material works and the consequences for high density recording. Firstly we must consider the control of grain size on the resulting noise generated by the medium on the written data. Figure 7.16 shows schematically examples of hypothetical media having different grain sizes. The arrows in Fig. 7.16(a) and

(b) show a point where the limit of a bit that was written would occur. Because the grains all contain a single magnetic domain they have their magnetic moment either oriented up or down and hence the sharpness of the transition will follow the line drawn along grain boundaries. Hence the width of the transitions will depend upon the grain size. Where small grains are used, the transition is intrinsically narrower than the case for the larger grains. Of course the schematically drawn larger grains could represent clusters of grains which are coupled together by an RKKY interaction and therefore all reverse together rather than as a single grain. Hence from simple geometric considerations it is clear that the narrowest transition widths will occur for systems with smaller grains and where there is control or elimination of the RKKY interaction. Since the level of silica in the films influences both of these parameters its inclusion has been critical to recent developments in the generation of ultra-high density perpendicular recording media.

It is of interest to observe exactly how the silica is grown within the grain structure. This is shown in Fig. 7.17 which shows both bright field and high angle annular dark field images of the silica at a grain boundary imaged via cross sectional TEM. This shows the remarkable use of the silica which extends down to the exchange break layer almost completely isolating each grain. The width of the silica in the grain boundaries is of the order of 1 nm being equivalent to about 3 atoms. In this sense even though the width of the disk is of the order of 95 nm that the advanced material science developed is resulting in atomic engineering of exchange decoupling.

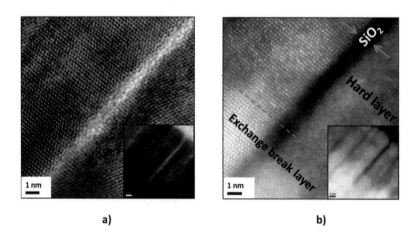

a)   b)

Fig. 7.17: Cross section TEM images of ECC media showing grain segregation by SiO_2 (Chureemart *et al.*, 2013).

We now turn to the effect of increasing levels of silica on the grain size distribution within the films. Figure 7.18 shows the effect of increasing levels of SiO_2 on the grain size and its distribution found in the films. Again these films and the measurements were made in our laboratories. For reasons of commercial confidentiality the exact level of SiO_2 used was not disclosed and simply listed as being high, medium and low. However the level of silicon shows not only a systematic decrease in the median grain size but also a systematic decrease in the width of the distribution which is important in terms of the resulting transition width shown schematically in Fig. 7.16. In the first image it is clear that the presence of significantly larger grains has been eliminated. This results from the fact that grain growth in granular films proceeds via a process of nucleation and growth followed by a third process known as Ostwald ripening whereby smaller grains fuse with larger ones thereby producing the largest grains that are commonly found in a range of particulate and granular materials both magnetic and otherwise. This results in the formation of a lognormal distribution function of grain sizes. Hence the effect on the transition width from the use of grain segregated media is clear.

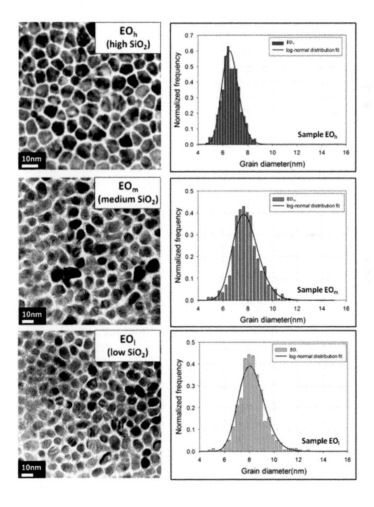

Fig. 7.18: TEM images and grain size distributions for co-sputtered media (Chureemart *et al.*, 2013).

For these samples we also studied the activation volume of reversal of the grains, the theory for which is discussed in Section 3.7. Figure 7.19 shows an example result of waiting time measurements discussed in Section 3.7 in a range of reverse fields around the coercivity. As required by eqn 3.40 lines of constant magnetisation are also shown and the value of the magnetisation at the intercepts of those lines is then used to calculate the variation of the reverse field with $ln(t)$ which then gives the activation volume at $M = 0$ via the equation. This together with the analysis of the TEM images allows a comparison between the magnetic activation volume at H_c and the physical volume of the particles. These data are shown in Table 7.1 below.

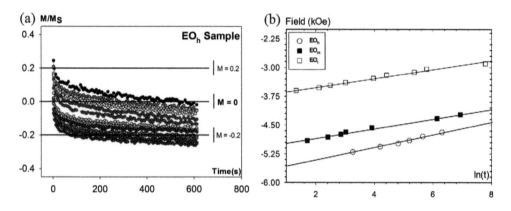

Fig. 7.19: Waiting time measurements for samples with different SiO_2 distributions (Chureemart *et al.*, 2013)

Table 7.1 Summary of physical properties of films with varying SiO_2 content.

Sample	SiO_2	D_m (nm) $\pm 1\%$	σ_D $\pm 1\%$	H_f $\pm 5\%$	V_{TEM} x10^{-19} (cc) $\pm 1\%$	V_{act} x10^{-19} (cc) $\pm 3\%$
EO_h	High	6.70	0.11	170	4.94	5.09
EO_m	Medium	7.81	0.13	127	6.52	7.32
EO_l	Low	8.16	0.13	114	7.21	8.20

This data shows a remarkable correlation between the value of the median volume taken from the TEM images and that obtained from the activation volume. In particular for the sample with the high level of silica, the value of the physical volume and the activation volume are in agreement within error. However for the medium level of SiO_2 the agreement is less good indicating the influence of the much reduced RKKY interaction which has not been completely eliminated. For the low level of SiO_2 again the correlation indicates the presence of some level of interaction. Table 7.2 shows a comparison of the key magnetic properties again along with the physical grain size. From this data the role of the exchange interactions is again clear. The most striking

Table 7.2 Summary of magnetic properties of films with varying SiO$_2$ content.

Sample	SiO$_2$	D$_m$ (nm) ±1%	H$_C$ (kOe) ±1%	slope at H$_C$ × 10^{-4} (Oe^{-1}) ±1%	M$_S$ (emu/cc)
EO$_h$	High	6.70	6.75	1.95	480
EO$_m$	Medium	7.81	5.41	2.27	450
EO$_l$	Low	8.16	3.97	3.00	450

observation is that a lower level of SiO$_2$ substantially reduces the coercivity due to the weak link effect discussed previously but the lower level of silicon also increases the squareness of the loop giving rise to a much narrower distribution of switching fields, represented by the slope of the hysteresis loop at the coercivity. Hence this data displays behaviour ranging from completely uncoupled very small grains to grains with an increasing level of RKKY interaction and the implications not only for the coercivity, but also the width of the switching field distribution. As discussed previously this implies that there is a compromise to be achieved with regard to the desired properties for high density recording.

Whilst these measurements were made as long ago as 2013 it is believed that very similar techniques remain in use at the time of writing (2022) and are included in the most modern hard disk materials currently in production associated with Heat Assisted Magnetic Recording (HAMR) discussed in Section 7.6.

7.4.2 Current Disk Structure

Much of the foregoing discussion of the structure of suitable disks for perpendicular recording are of course idealised structures and not necessarily those that are in current use. It is difficult to find information on the actual structures used because the major hard drive manufacturers do not put such information into the public domain. However a typical disk structure and a very clear explanation of the role of the various layers is shown in Fig. 7.20.

As can be seen the disk structure consists of a substrate which at that time would still be glass or an AlMg alloy with an incredibly low rms roughness which occurs naturally for glass but which requires the NiP smoothing layer for the case of an AlMg alloy. On top of that there is an adhesion layer which is an alloy of AlTi which is followed by a tri-layer structure to form of a synthetic antiferromagnet (SAF) discussed in Section 5.3. This structure acts as the SUL so as to produce an image of the single pole head and hence focus the magnetic field. The SAF is used because whilst it is composed of two magnetically very soft layers, due to the presence of Zr in the alloy, when not exposed to a field from the single pole head the SAF will have a magnetic structure where both layers will be saturated in opposite directions. In this way there should be few or even no domains in the SAF and hence there will be no spurious signals from domain walls when data is being read. In this case the company had used a NiWCr seed layer which presumably forms a good lattice match to the basal plane of a layer of Ru which will then grow with the c-axis perpendicular to the plane. The purpose of that layer is to improve the orientation of the recording layer so as to achieve almost an ideal Stoner-Wohlfarth structure which will store the data. As indicated,

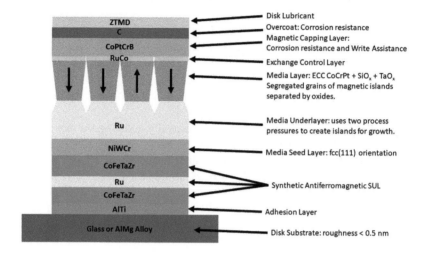

Fig. 7.20: Current conventional perpendicular media stack with segregated magnetic grains.

the media layer itself is an ECC media CoPtCr magnetic material but in this case a combination of SiO_x and TaO_x oxide layer has been used. Presumably in each case the value of x will be close to 2 so as to provide insulation between the magnetic grains hence restricting the RKKY interaction. There is then an additional continuous layer of CoPtCrB which will be a soft magnetic material previously discussed and known as a CGC (Continuous Granular Layer) which will act like a keeper to help to protect the stored bits of information from demagnetising effects. Again an exchange break layer is included between the recording and CGC layer to prevent indirect RKKY coupling between the grains in the recording layer.

As discussed a carbon overcoat and a disk lubricant are included but in this case it does not appear to be a perfluorinated polyether. Hence from this structure it can be seen that a number of innovations made in magnetic nanotechnology over the years have all been deployed together to produce a viable disk for perpendicular magnetic recording.

7.5   The Read-Write Head

For a hard drive or any other magnetically based information storage system to operate correctly requires careful integration of the read and write heads which have to be designed to ensure that they are compatible with the disk structure. It is also necessary to have an electronic system such that the signals emanating from the read head can be interpreted correctly to give a string of 0s and 1s. The signal analysis system is known as the channel.

Precautions have to be taken to ensure that an individual bit is read without interference from any neighbouring bits. Spurious signals from neighbouring bits produce a signal known as cross talk. Other precautions also have to be taken to limit any noise being generated in the material due to stray magnetic fields coming from the individual layers in the device. It is not the purpose of this textbook to describe the engineering of a hard drive system but to focus on the behaviour of the particulate and granular materials of which the vital components i.e. the disk and head are composed. Hence other than explaining the reasons for the choice of materials and for the aspects of the design in broad terms, these will not be discussed.

We have already discussed the nature of the write head in Section 7.3 and in particular the fact that it is a single pole head. Such a device shown schematically in Fig. 7.12 in principle appears very simple. However this device is actually produced by complex thin film deposition combined with reactive ion etching (RIE) to produce the net shape required and possibly with ion beam milling to narrow the width of the single pole head. The production of this device is a significant challenge because of the requirement to deposit, in thin film form, a coil around the bridge in the quasi-horseshoe shape of the device. Of necessity the write head is significantly larger than the read head to allow for the various components to be produced and as shown in Fig. 7.12, the read head is deposited directly onto the back of the return pole of the write head.

The read head itself is a highly complex magnetic and engineering structure to produce the required signals free of cross talk and noise generation. It had been claimed with good cause, that the read head used in a conventional perpendicular recording hard drive is perhaps the most complex miniaturised device ever produced.

All read heads for perpendicular recording are based on a tunnelling magneto-resistive (TMR) structure discussed in Section 5.4 but with sophisticated additional components to enhance the performance and satisfy the engineering specification. A cross sectional TEM image of a read head stack is shown in Fig. 7.21(a). Of course for perpendicular magnetic recording the layers in the stack must lie perpendicular to the plane of the disk and hence this image would be the view of the device that would be seen when looking up from the surface of the disk. An enlarged view of the tunnel junction is shown in Fig. 7.21(b).

As can be seen from the schematic diagram in Fig. 7.21(c) is a description of the various layers but not drawn to scale. The first key feature are the shields which appear at the top and bottom of the thin film stack. These shields exist to prevent cross talk from neighbouring bits along the track which is being read. They are typically composed of an alloy NiFeCr but the exact composition is not in the public domain. Suffice to say that a simple alloy of NiFe with a composition of 80% Ni and 20% Fe, known as permalloy, and is one of the standard ultrasoft magnetic materials i.e. having a very low coercivity. It is believed that the Cr is included in the composition to tune the magnetic properties and also to enhance corrosion resistance. This layer acts as the shield and also the bottom electrode as this is a CPP device.

Working up from the bottom of the device as shown in Fig. 7.21(c) the next layer is composed of IrMnCr which is the AF layer (8 nm) that generates the exchange bias for the pinned layer of $Co_{40}Fe_{60}$ in the TMR stack. In some designs this layer of CoFe forms part of a SAF structure to enhance the pinning. The AF layer can be set perpendicular to the plane of the diagram and the SAF used to enhance the thermal stability of the AF grains as discussed in Chapter 4.

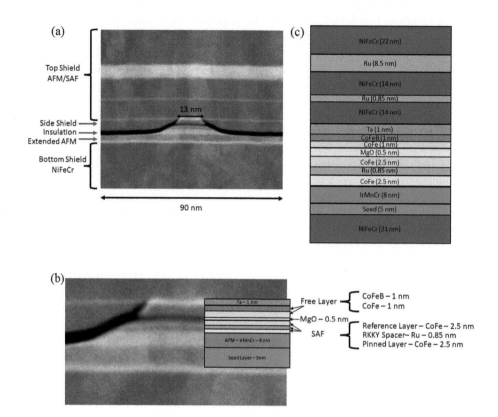

Fig. 7.21: (a) Cross section TEM image of a read head device (b) Enlarged view of the tunnel junction, (c) Schematic of the layers in the device.

The insulation layers and the side shield structure serve to protect the TMR device from strong electrical effects, eg static, and from cross-talk from neighbouring tracks. It is believed that this device was designed to read bits on tracks only 23 nm wide with bit lengths of 10-20 nm.

The TMR device is then formed by the 2.5 nm CoFe layer separated by 0.9 nm of MgO, which is only 3 atoms thick, and must be crystalline and grown epitaxially to the ferromagnetic layers. This is achieved by having a free layer consisting of 1 nm of CoFeB and 1 nm of CoFe. The CoFeB deposits in an amorphous phase which on annealing crystallises at a low temperature and induces epitaxial crystallisation of the MgO. The two high magnetisation CoFe layers ensure the maximum TMR as given by eqn 5.6.

The remainder of the layers above the TMR structure form the top shield for the head. It is not known why a complex, double SAF is used but presumably this is to ensure uniform absorption of stray flux from neighbouring bits.

Given the complexity of this device, the film thicknesses involved and the dimensions, the description of it as the most advanced device ever mass produced is probably justified. This is especially so when one considers that over 100,000 such devices are produced and 100% tested on each 8" (20 cm) wafer.

7.6 Heat Assisted Magnetic Recording

7.6.1 Limits to Conventional Recording

Over the years there have been a number of usually, theoretical papers that claimed that conventional magnetic recording was reaching its limits. In fact there was even one paper that suggested that a data density of 1 Gbit/in^2 was technically impossible. That and other similar papers were clearly wrong. Using ECC media based perpendicular magnetic recording disks with a top layer of a continuous granular coupled layer to help stabilise the data, enabled Seagate Corporation to produce disk drives with a density of 1.4 Tbit/in^2. However at some point there must be a limit as to how small bits of information can be made and remain readable and thermally stable.

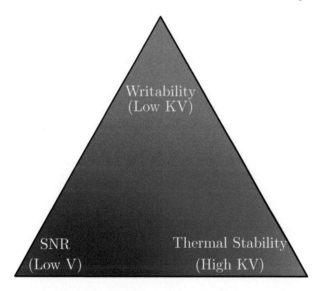

Fig. 7.22: The recording trilemma to balance writability, SNR and thermal stability, all of which require conflicting properties.

This conundrum is best represented by what came to be known as the recording trilemma shown in Fig. 7.22 above. The recording trilemma shows the three essential ingredients for data storage: the first of these is that the medium must be writable. That means that the anisotropy energy barrier given by KV must be sufficiently low to enable the field from a single pole head to switch the magnetisation of the grains. That field has an intrinsic limit imposed by materials with a sufficiently high saturation magnetisation to generate the necessary flux density and the ability to shape the end of the single pole head using techniques such as ion beam milling. The material of choice has to be the alloy $Co_{40}Fe_{60}$ which is known as permendur or cobalt-iron, and has a saturation magnetisation of 24 kG which is the highest known magnetisation of any element or alloy. Hence the field cannot be increased by using any other material. Similarly shaping of the single pole head has now also reached its limits and as we have seen in Section 7.4, it is only the ingenuity of the engineers and material scientists working on disk structures and who developed the concept of ECC media media, that it has been found possible, albeit indirectly, to switch a storage layer with a coercivity of significantly greater than 4 kOe. Hence in terms of writability we have already reached, if not exceeded, the limit of what is possible.

The second requirement is that the stored data and hence the magnetisation of the individual grains must be thermally stable. The thermal stability of the magnetisation of granular materials was discussed in detail in Section 3.5 and fundamentally the requirement is to have a very high energy barrier given by KV. This is particularly true for grains oriented perpendicular to the plane of the thin film because there will always be the inevitable demagnetising field given by $H_d=-4\pi M_s$ which will reduce the barrier to a thermally activated reversal depending on the value of the anisotropy field H_K which itself depends on the inverse of M_s ($H_K = 2K/M_s$). Even some of the granular structures discussed with regard to ECC media media were found to be insufficiently thermally stable to be used when the grain size was below 8 nm. Hence once again we find ourselves at the limit of what can be achieved with conventional materials.

The third element of the trilemma is the requirement for a low signal to noise ratio which as shown in Fig. 7.16 requires the grains to have the smallest possible size. Obviously we can reduce the grain size further but then the thermal stability becomes compromised.

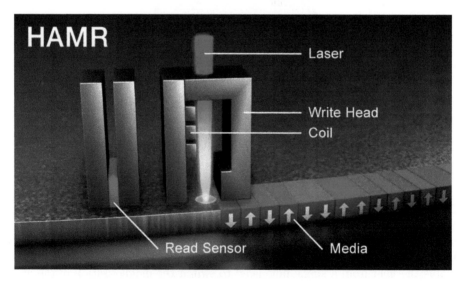

Fig. 7.23: Schematic illustration of the operation of HAMR media.

It was therefore realised by 2010 that for the continued advancement of magnetic storage technology a new paradigm was required. Even before that date there was already discussion among research scientists about solving the magnetic recording trilemma by the use of an additional factor so that a material with very high anisotropy could have its anisotropy reduced by the use of a heat pulse which would make the media writable and, after rapid cooling, an increase in the anisotropy would result that would render the written bits of information thermally stable. If a material with sufficiently high anisotropy could be produced in the required format then that in turn would allow a further reduction in the grain size and hence the signal to noise ratio. That is the principle of heat assisted magnetic recording commonly known as HAMR. The basis of a HAMR system is shown in Fig. 7.23.

7.6.2   Heat Assisted Magnetic Recording (HAMR) Disk Structure

Despite the ingenuity mainly of the industrial engineers working in the major magnetic recording companies, it became clear shortly after 2010 that even with the paradigm shift to perpendicular magnetic recording and concepts such as ECC media media and the use of a CGC layer to increase data density, that conventional magnetic recording would eventually run out of steam. There were two principal reasons for this, the first being the fact that a conventional write head even with the assistance of a soft underlayer, could not generate a field sufficient to switch materials with very high anisotropy and the second issue was that in order to record data at ultra-high densities would necessitate the use of ever smaller grains as discussed in Section 7.4 to allow for further narrowing of the transition width.

By 2015 the record for conventional magnetic recording was 1.2×10^{12} bits/in^2. Also by that time initial work on the concept of HAMR had already begun principally by Seagate Corporation, who had demonstrated a data density of 1.4 Tb/in^2 demonstrating the potential of a heat assisted system to extend the data density beyond what was capable with conventional recording and potentially out to densities of 4 Tb/in^2 and possibly beyond.

The principle of HAMR is that the use of a material with very large magnetocrystalline anisotropy results in a temperature variation of the anisotropy as shown in Fig. 7.24. The material with the highest known anisotropy at that time was an alloy of FePt. This alloy, when grown in the L1$_0$ phase, can have a uniaxial anisotropy constant in thin film form as high as 4.5×10^7 ergs/cm^3 which is only 25% below that of an FePt single crystal for which the value is equal to 6.6×10^7 ergs/cc (Weller *et al.*, 2016).

Anisotropies of this magnitude have been demonstrated to give coercivities in thin film form as high as 50 kOe or in the SI system, 5 T. Obviously such materials cannot be reversed using a magnetic field other than that generated by a superconducting magnet. However FePt has a Curie temperature as a bulk material of 750 K or 477°C. Applying such a temperature to a disk structure is likely to cause significant damage not only by diffusion of layers, but also by distorting the substrate itself. Hence a number of additives were used in an attempt to lower T_C. This was done by the addition of elements such as Cu and Ag which at a level of about 10 at. % can reduce the Curie temperature by $\sim$100 K.

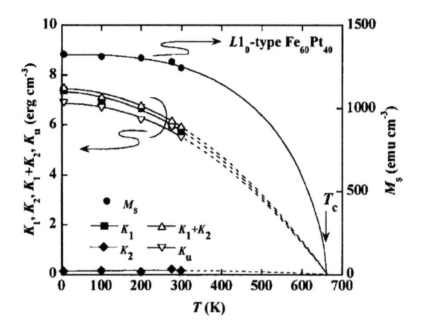

Fig. 7.24: Temperature dependence of the magnetocrystalline anisotropy for FePt. Reprinted from Inoue *et al.*, Appl. Phys. Lett. 88 102503 (2006), with the permission of AIP Publishing.

From Fig. 7.24 it is now clear how the principle of a HAMR recording system is based. Basically the rapid drop off of the coercivity resulting from the reduction in the anisotropy means that data could be written with a conventional single pole recording head at a temperature somewhat above the actual Curie point as the anisotropy decreases more rapidly than the magnetisation. It is then necessary to induce rapid cooling so that the data is then stored at room temperature where the anisotropy is close to its maximum value. The use of a technique such as this would allow for the growth of grains with good thermal stability against thermal loss of data that could be as small as 4 nm hence allowing for a significant increase in data density whereas the conventional grain size for conventional perpendicular recording is around 8 nm to achieve thermal stability.

Of course the growth of $L1_0$ FePtX alloy grains with perpendicular orientation and at a very small size provided a whole new set of challenges for those scientists and engineers involved in the growth of such films. Additionally there was still a requirement to induce grain segregation into the system so as to obviate the effects of RKKY coupling. However this need is reduced due to the high anisotropy which will oppose the RKKY coupling.

A further difficulty occurs in that alloys of FePt when deposited by conventional DC magnetron sputtering do not grow in a chemically ordered state unless the film is subsequently annealed to very high temperatures. This problem was reduced by the growth of this layer at an elevated temperature. However growth temperatures to produce higher crystalline and chemically ordered thin films was found to be as high as 650°C. It was also found that the growth of chemically ordered and crystalline grains was best achieved by the use of a MgO seed layer which then results in the formation of well oriented FePt (002). This resulted in a FePt lattice parameter ratio $c/a \sim 0.96$ and a high chemical order parameter $S > 0.9$ which gave the high value of the anisotropy for a single film quoted above. Given the level of the anisotropy of these alloys the switching field distribution now becomes almost entirely dependent upon the distribution of the anisotropy field H_K. As discussed in Section 2.4 the distribution of H_K is most probably due to slight variations in the easy axis directions where a shift of 10% from the perfecly aligned case reduces H_K by 30%. However with the use of the correct seed layers a variation in the anisotropy constant $\Delta H_K / H_K \sim 10\%$ but in this case that variation is equivalent to a variation in the anisotropy field of the order of 10-12 kOe with the value of H_K then being $\sim$ between 10-11 T at room temperature. A further problem with the use of MgO seed layers is of course that such an oxide is insulating and hence the requirement for rapid cooling of the written disk becomes more difficult (Weller *et al.*, 2016).

It was also found that the use of SiO_2 as the grain segregant also inhibited the crystallisation of the FePt. Following extensive studies by a wide range of material scientists but notably by the group of K. Hono at the NIMS Institute in Tsukuba, Japan it was found that for example, the inclusion of carbon or boron nitride (BN) actually promoted the crystallisation of the FePt but also resulted in segregation layers based upon FeC. Hence a significant effort in material science was found essential to basically reinvent a magnetic recording disk for the purposes of HAMR (Bolyachkin *et al.*, 2022).

Again it was suggested and proved entirely feasible, that a lower temperature heat pulse could be used if the FePtX-Y layer, where X represents the element included to lower T_C and Y is the grain segregant, were overgrown with a layer of the alloy CoPt which grows with a similar crystal structure. CoPt has a significantly higher Curie temperature of ~ 840 K and again similar additives and segregants were used as in the FePt layer. This means that the CoPt layer can be switched at a lower temperature and then via exchange coupling through the same mechanism as was used in conventional ECC media was then used to switch the FePt layer (Victora *et al.*, 2017).

Remarkably it was found to be beneficial and avoided damage during the necessary high temperature growth of the high anisotropy layers, to not include a soft underlayer (SUL). In the case of HAMR media the bit size is controlled by the spot size of the laser which is used within the write head to heat the disk. Hence by detailed control of the spot size it was possible to dispose of the additional heat sink. Hence the structure of the disk was greatly simplified as shown schematically in Fig. 7.25.

Of course it should be noted that the majority of the development work on media for HAMR has been undertaken within the major recording companies research labs and at massive expense by a large number of scientists and engineers. Hence very little is published in the open literature and much of what has been described has been inferred from the few published papers that are available, very few of which have been published in the recent past.

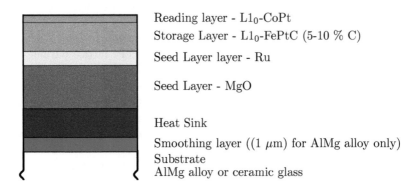

Reading layer - L1$_0$-CoPt

Storage Layer - L1$_0$-FePtC (5-10 % C)

Seed Layer layer - Ru

Seed Layer - MgO

Heat Sink

Smoothing layer ((1 μm) for AlMg alloy only)
Substrate
AlMg alloy or ceramic glass

Fig. 7.25: Schematic disk-structure of HAMR.

7.6.3   HAMR Recording Head

The read/write head for the HAMR system is now basically the same as that for a perpendicular recording medium. In particular the read head is virtually unaltered other than by some scaling so that the read head is able to detect ever smaller bits. However magnetically the write head is again exactly the same with a single pole head. The major advance that has been made is the inclusion of a laser which has an assembly such that the spot is generated using near field optics so as to breach the normal diffraction limit and a near field transducer to deliver the spot with the correct optical profile to the surface of the disk. In general a spot of laser light produced in this way has a Gaussian intensity profile but of course spots in a recording head are produced via a semiconductor laser so further optical processing to produce a circular spot is also required.

A schematic of how the read/write head is produced is shown in Fig. 7.23 but the detail of the optical system constitutes the majority of the device and is not magnetic in nature. Suffice to say that extreme measures have been taken to distort the nature of the intensity distribution in the spot so as to only have the narrowest possible region where the temperature is sufficient to reverse the medium. Again the major manufacturers have published very few if any detail of this photonic device and so little of the practicalities of how the device is produced or works is publicly available. Of course the topic of this text does not include photonics so details of such head designs are beyond the scope of this work.

7.6.4   The Future of Disk Recording

In 2017 Seagate announced that they had produced the first hard drives based on HAMR and again it is believed that they shipped demonstration devices to potential customers in late 2018 or 2019. Seagate have announced that they now have many tens of thousands of hard drives based on HAMR with customers for integration and durability tests. Such drives are simple to integrate and operate in a similar manner to any traditional drive and have passed qualification tests with the end users. It has been announced that they shipped pilot volumes in 2020. Because of the extraordinary high data density it is now possible to buy a single hard drive with a capacity of 20 TB and that in the next few years a similar annual density growth of HAMR media data density of around 30% per annum will be achieved.

It has been predicted that by 2025 the amount of stored data will grow to a 163 Zettabytes (ZB) which is 10 times the amount of data being stored in 2017. It is expected that hard drives will be responsible for 70% of that storage mainly in data centres commonly known as cloud computing. We all know that almost all PCs produced today use solid state electrostatic memories for local storage and for mass storage everything else goes to the cloud and 70% of the cloud is stored magnetically. Seagate have also claimed that they have working drives for passing 2 Tbit/in^2 and the industry projection is that this will double in the next few years.

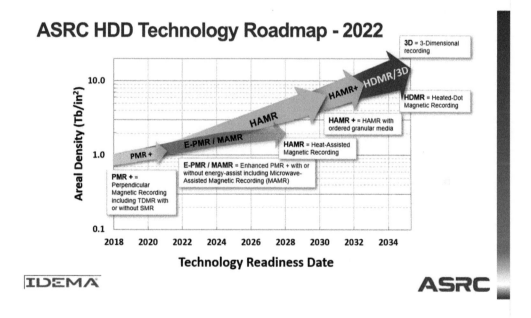

Fig. 7.26: ASRC technology roadmap (Courtesy of ASRC, 2022).

The magnetic recording industry and others involved, have an industry association known as the Advanced Storage Research Consortium or ASRC through which they cooperate on matters commercial and to a limited degree technical where common interests are involved. Figure 7.26 shows their current thinking about the future of magnetic recording technology up to 2034. This shows that they believe that conventional HAMR systems will run out of capacity at a data density of about 4-5 Tbit/in^2

and thereafter the only way to further advance storage densities would be to go to some sort of nanodot pattern system. However given that the current bit size on a HAMR system is between 10-20 nm and the track width is now down to 23 nm it is somehow hard to envisage how a patterning system could be developed that would make a significant advance economic.

It is not often realised by those involved in a purely scientific endeavour that on a conventional or even a HAMR disk it is necessary to first write a servo track which keeps the head centred on the track being read. As well as taking a lot of time to write, this track which is magnetic in nature and has a chevron pattern alongside the linear orientation of the bits to be read, takes up approximately half the capacity of the disk. With a patterned structure it may be then possible to servo the head off the edge of each cell in the pattern and in that way the data stored on a disk could in principle be increased by a factor 2 but no more, unless the cell size was reduced to below 10 nm in each direction. The cost implications in undertaking such an exercise would probably make such a system uneconomic but that being said, 10 years ago the same was said about HAMR.

8
Magnetic Random Access Memory (MRAM)

8.1 Principles and Early Designs

In principle the same concepts of magnetic information storage as used in hard disc drive technology could similarly be applied to creating a solid-state magnetic memory. There are obvious disadvantages to the use of magnetics to store information because, based on Faraday's law, it is necessary to move the storage medium so as to generate a signal which can be read and interpreted. This clearly makes the process quite slow and also involves the use of a significant amount of energy. Based on the demand for information storage this has significant implications for priorities such as reducing the emission of carbon dioxide into the atmosphere as many of the data warehouses currently in use for cloud storage involve significant energy inputs to rotate the discs on a hard drive and then, because that energy dissipates heat, significant cooling is also required. In the medium term given the implications for global warming, this is highly undesirable.

There also exists a further problem when one considers storage on hard drives with a rotating disk. Again due to Faraday's law, if a disk is rotating at a constant rate which these days is typically 7500 rpm then obviously the linear speed near the edge of the disk is significantly faster than that near its centre. Hence the output signal for transitions near the edge of the disk is much higher than those near its centre.

From a technological point of view a small magnetic element comprising an MRAM cell can be switched a thousand times faster than is the case for a silicon based device due to the value of the attempt frequency (f_0) being of the order of 10^9 Hz which corresponds to a switching time of 1 ns. In practice switching times are significantly longer than 1 ns but switching times of 5-10 ns are predicted to be possible and recently have been achieved. There are other issues with Si based storage technologies such as Flash Memory in that the durability under constant switching of such systems can be limited whereas it is possible to switch a magnetic entity potentially an infinite number of times. This has the impact of making such a potential solid state magnetic memory far more durable than is the case with other technologies.

The concept of a solid state MRAM technology was first proposed as long ago as 1972 by Schwee. The original concept proposed by Schwee was based on a solid state magnetic memory consisting of the storage of information in domains on a thin film. During the late 1970s and 1980s there was a huge amount of interest in what came to be known as bubble domains which were based upon this original idea. Some success was achieved and a commercial device was launched by Texas Instruments in 1977 (Schuyten, 1978). The reasons for this are based around the typical width of a domain and the associated domain walls which generally lie in the micron range. However more recently Parkin and co-workers proposed what became known as racetrack memory where rather than storing information in a bulk thin film wires of a suitable thin film which were lithographically defined allowed for the possibility of producing significant smaller entities (Parkin *et al.*, 2008; Hayashi *et al.*, 2008). In the case of the system proposed by Parkin *et al.* the position of the stored bits of information were to be detected by GMR or TMR. The wires were fabricated with notches on their sides which would act as domain wall pins (Parkin *et al.*, 2008; Hayashi *et al.*, 2008).

Following the discovery of GMR and subsequently TMR in thin films by Jullière in 1975 and by Maekawa and Gafvert in 1982 a clear mechanism for reading the magnetic orientation of a layer in a small element became clear as the GMR or TMR effect was known to generate a significant signal that could be read easily. Figure 8.1 shows schematically the basic concept of an MRAM device consisting of cells deposited on top of bit lines and word lines that enabled the writing and reading of the stored information. Also shown in the diagram is how these early designs were developed subsequently through a longitudinal configuration employing an exchange bias structure in a GMR configuration through to a similar structure for a TMR device and a device with the orientation perpendicular to the plane of the layers i.e. a CPP structure (Saito, 2021).

Based on these configurations where arrays of MR junctions sandwiched between bit and word lines preferably with vertical alignment, for considerations similar to those discussed in Chapter 7 for perpendicular magnetic recording, it is clear that the configurations can be switched by sending a pulse current for writing and detected by measuring the MR signal under a sensing current. This allows for full electrical operation unlike conventional magnetic recording which requires mechanical movement of the storage medium and the read and write head which must scan across tracks on a rigid disk. The first MRAM device incorporating a GMR junction using an Fe-Ni-Co alloy was made by Granley *et al.* in 1991 (Granley *et al.*, 1991). Because of its greater sensitivity it was rapidly realised that a TMR junction based device would have far greater sensitivity and these were produced in 1996 (Wang and Nakamura, 1996). The subsequent historical development mainly in industrial labs, is summarised in Table 8.1 which shows a calendar of key early developments of MRAM technology.

MRAM has been developed as a universal memory to fulfil the performance gap between temporary (DRAM) and permanent storage in a computer which these days is currently based on Flash Memory, which is a form of storage based on silicon technology. The current computational hierarchy has a central processing unit (CPU) which operates at GHz clock speeds where the data is fed from level caches with a latency of nanoseconds, i.e. Static Random Access Memory (SRAM). The level caches are fed by temporary storage with a latency of 10 ns and a larger capacity, i.e. Dynamic Random Access Memory (DRAM). These volatile memories store data in a flip-flop circuit and a tiny capacitor for SRAM and DRAM respectively. For DRAM the data needs to be refreshed. Additionally the temporary storage is fed by permanent storage with a latency of $\sim$100 ns and a much larger capacity such as that available with Flash

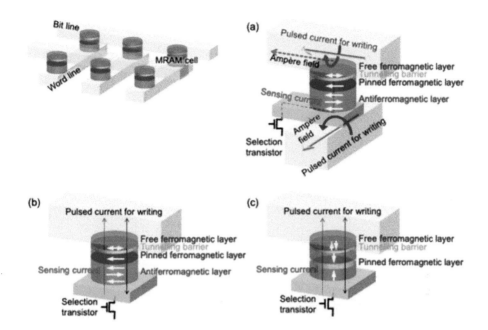

Fig. 8.1: Schematic diagrams of MRAM with the schematic cell configurations of the (a) first, (b) second and (c) third generations. Used with permission of Saito, from Magnetic random access memory in A. Yamaguchi, A. Hirohata and B. Stadler (Eds.), Nanomagnetic Materials: Fabrication, Characterization and Applications (Elsevier, Amsterdam, Netherlands) 486-501 (2021); permission conveyed through Copyright Clearance Center, Inc.

Memory. This creates a performance gap between the temporary and permanent storage. A comparison between MRAM and other solid state memory systems is shown in Table 8.2.

In order to fill the gap, a universal memory based on MRAM has now been under development for approximately 20 years. The system consists of one selection transistor and one memory cell e.g. a capacitor, a resistor and a magnetic tunnel junction. A ferro-electric capacitor, phase-change resistance or an MTJ can be used for Fe RAM, P RAM, Re RAM and MRAM respectively. Flash Memory is a form of electrically erasable programmable read only memory (EPROM) and uses an extra floating gate on a transistor to store data, a large section of which can be erased by applying a voltage pulse which can induce permanent damage to the memory. It is this intrinsic defect that leads to the limited durability and long write times in such systems. Table 8.2 shows a direct comparison of the performance of these various types of memory. As can be seen in the table, MRAM is the preferred candidate for a universal memory by achieving non-volatility with full electrical operation. It also has nanosecond latency and a high density of data per unit area. There is also low power consumption for operation.

Table 8.1 Historical development of MRAM.

Year	Authors	Storage Capacity	Notable Features
1972	Schwee		First proposal of MRAM
1991	Granley *et al.*		GMR device of Fe-Ni-Co
1996	Wang and Nakamura		Demonstration of TMR MRAM
1997	Daughton		Demonstration of TMR MRAM
1999	IBM, Motorola	1-kbit, 512-kbit	
2003-2005	Motorola	1-Mbit - 4-Mbit	
2003-2006	NEC-Toshiba	1-Mbit - 16-Mbit	
2004	DeBrosse *et al.*		Latency down to ~5 ns for read/write
2007	Freescale Semiconductor	4-Mbit	Commercialised with operating range from (-40 - 105)°C
2010	Everspin Technologies	16-Mbit	35 ns speed.

Table 8.2 Comparison between MRAM and major memories (Hirohata and Takanashi, 2014; https://knowen.org/nodes/27127; https://www.extremetech.com/computing/163058-reram-the-new-memory-tech-that-will-eventually-replace-nand-flash-finally-comes-to-market; http://www.thelec.net/news/articleView.html?idxno=2428).

	SRAM	DRAM	Flash	FeRAM	PRAM	ReRAM	MRAM (1st)	MRAM (2nd)	MRAM (3rd)
Data Retention	Volatile	Non-volatile	Non-volatile	Non-volatile	Non-volatile	Non-volatile	Non-volatile	Non-volatile	Non-volatile
Read time (ns)	$0.2\sim2$	10	$30\sim5\times10^2$	2.3	115	100	300 (GMR) <60 (TMR)	105	
Write time (ns)	$0.2\sim2$	10	$10^5\sim10^8$	2.3	50	6×10^3	< 10	170	
Endurance	10^{16}	$>10^{15}$	$>10^6$	10^{14}	10^6	10^6	$>10^{15}$	$>10^{10}$	$>10^{15}$
Fabrication rules F (nm)	5	45	20	130	20	130	90	40	28
Cell size [F^2]	92 (6T)	6 (1T1C)	$5.4\sim18$ (1T1C)	15 (1T1C)	4 (1D1R)	8 (1T1R)	27 (2T1MTJ)	8 (1T1MTJ)	
Cell size (μ^2)	$\sim2\times10^{-2}$	$\sim2\times10^{-3}$	$\sim2\times10^{-2}$				3.87×10^{-1}	1.56×10^{-1}	3.96×10^{-2}
Bit density (Mbit/cm^2)	64	$\sim2.5\times10^5$	$\sim2.5\times10^3$				98	$\sim2.56\times10^2$	$\sim9.75.5\times10^2$
Chip capacity (Gbit)	$12\sim16$	$12\sim16$	>1	<0.01			6.4×10^{-2}	2.56×10^{-1}	1
Programme energy (pJ/bit)	$30\sim1\times10^2$	5+refresh	1×10^2	0.37		$24\sim1.5\times10^3$	120	1	
Soft error hardness	No	Yes	Yes	Yes	Yes	Yes	Yes	No	
Stackable				No	Yes	Yes			
Multi-level cell	No	Yes	Yes	No	No				
Process cost			Low bit cost	High temperature process			Room temperature process		

8.2 Initial Implementation

8.2.1 First Generation Toggle-MRAM

In principle the cross-point architecture shown in Fig. 8.1 allows for a very high aerial density. Initially it was proposed that each MRAM cell would be reversed by an Ampere field generated by a pulse current for writing as shown in the figure. It was also a requirement that magnetic shield layers had to be installed above the bit line and below the word line to confine the fields generated and minimise cross-talk to neighbouring cells in a similar manner to the shields used in read heads for conventional magnetic recording described in Section 7.5. The summation of the two Ampere fields generated by the bit and word lines can be aligned along a diagonal direction at the cross point of the selected bit and word lines, which is parallel to the easy axis of the Magnetic Tunnel Junction (MTJ) and MRAM cell. The MTJ was generally patterned to an elliptical shape where the long axis produced a shape anisotropy in the cell to enhance the coercivity. The switching field $H_c(\theta)$ is given by

$$\frac{H_c(\theta)}{H_K} = \left(cos^{2/3}\theta + sin^{2/3}\theta\right)^{-3/2} \tag{8.1}$$

where H_K and θ are the uniaxial anisotropy field and the angle between the generated Ampere field and the easy axis of the MTJ. This gives rise to Stoner-Wohlfarth behaviour which determines the switching field for the MRAM writing operation.

Of course to achieve Stoner-Wohlfarth behaviour it is necessary that the entire cell be a single domain entity. Whilst there is no clear definition of the exact transition from multi domain to single domain, it is generally found that it is the smallest dimension of an entity that determines its domain state. Hence in the case of an MRAM cell such as that shown in Fig. 8.1, it will be required that the free layer be relatively thin which should ensure a single domain state. However it remains the case that the cell size would be required to be of the order of 1 μm or less otherwise significant incoherent reversal discussed in Section 2.5 may well occur thereby limiting the coercivity that can be achieved.

For reliable cell selection with an increased margin for switching, toggle switching was proposed and implemented to circumvent the half selection in an MRAM cell, see Fig. 8.2 (Engel *et al.*, 2005). By sending the pulse currents in the bit and word lines with some delay, the magnetisation of the free layer in the GMR or TMR structure can be switched via the step along the hard axis which is aligned in the MTJ perpendicular to the easy axis. The corresponding toggle field (H_T) for switching is given by eqn 8.2

$$H_T = (H_K H_S)^{1/2} \tag{8.2}$$

where H_S is the magnetic field required to saturate the magnetisation of the free layer. This toggle MRAM system was commercialised by Freescale in 2007. This company is now part of Everspin Technologies.

As is the case with conventional magnetic recording it is required that MRAM bits exhibit good thermal stability. In principle the requirement is that the anisotropy energy $KV/kT > 100$ as is the case in conventional disk recording. However it should be noted that this criteria whilst being valid for the in-plane case, will also need to take

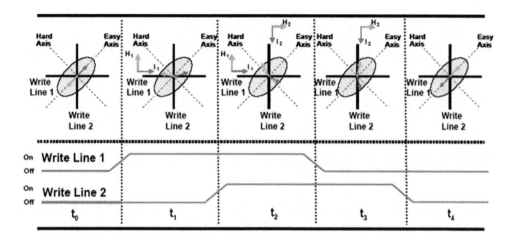

Fig. 8.2: Schematic diagram of the toggle switching. Pulsed currents are applied in a sequence that rotates the synthetic antiferromagnet by 180° to the opposite magnetisation. Reprinted from Magnetic tunnel junction based magnetoresistive random access memory M. Johnson (Ed.), Magnetoelectronics, Akerman *et al.* 231-272, Copyright (2004), with permission from Elsevier

account of the perpendicular demagnetising field when the orientation is perpendicular to the plane of the film. This requirement is also necessary to avoid half selection of the switching of a cell. As discussed in Section 5.3, a synthetic antiferromagnet (SAF) consisting of two ferromagnetic layers sandwiched between a non-magnetic layer creates an anti-parallel configuration which has been used as the pinned ferromagnetic layer to maintain the required high thermal stability (Inomata *et al.*, 2003). By aligning an elliptical MTJ along the diagonal of the bit and word lines the magnetisation of the free layer can be reversed by an Ampere field. For commercial MRAM devices reliable readout requires one additional transistor for selecting the MRAM cell required.

8.2.2   Second Generation STT-MRAM

One of the issues with the entire concept of MRAM has always been the power consumption to switch a magnetic element using an ampere generated magnetic field. As shown in Fig. 8.3, as the cell size is reduced the required ampere field to switch such an element increases dramatically. This increases the power consumption significantly as the data density is increased via the use of smaller elements. This effect which of course arises as the element size approaches the single domain size and hence magnetisation reversal is no longer via some form of domain wall process, appeared to represent an insurmountable difficulty for high density MRAM technology.

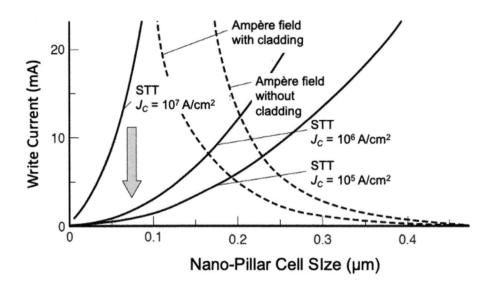

Fig. 8.3: Required writing currents for several techniques dependent upon cell size (Nakamura *et al.*, 2006).

Following the prediction and verification that a small element could have its moment reversed via a spin transfer torque mechanism discussed in Section 5.5, this led to a dramatic set of developments associated with what is now known as STT-MRAM. Again as shown in Fig. 8.3 (Nakamura *et al.*, 2006) as the element size was reduced the required equivalent current density to switch smaller elements decreased in an exactly opposite way than was the case with an ampere field. The concept of spin transfer torque was proposed and verified in 1996 and at this stage the capability of lithographic techniques and particularly reactive ion etching where a plasma is struck in a medium that contains reactive ions, meant that the potential for a high density MRAM storage system was now increased dramatically. Spin transfer torque can be quantitatively described as

$$STT = \alpha \mathbf{m} \times \frac{d\mathbf{m}}{dt} \tag{8.3}$$

where $\mathbf{m}$ is the unit vector of the magnetisation. By adding this term to the conventional Landau-Lifshitz-Gilbert (LLG equation) the magnetisation dynamics in an MTJ is given by

$$\frac{d\mathbf{M}}{dt} = -\gamma \mathbf{m} \times \mathbf{H}_{eff} - \frac{\gamma}{d}\mathbf{m} \times (\mathbf{m} \times \Delta \mathbf{J}_s) + \alpha \mathbf{m} \times \frac{d\mathbf{m}}{dt} \tag{8.4}$$

In this equation the second term is the relaxation term governed by the Gilbert damping constant α which describes the coupling of the magnetisation precession to the phonon bath. This term increases with increasing temperature. Conduction electrons are scattered by localised spins during the spin relaxation process releasing the corresponding momentum into the lattice via the spin-orbit interaction at high temperatures. This results in spin relaxation, i.e., α is proportional to the resistivity of the system (ρ) (resistivity like) while α is proportional to the conductivity of the system σ (conductivity like at low temperature). In eqn 8.4 the STT is antiparallel to the Gilbert damping torque indicating that the increase in the spin current density (J_s) reduces the relaxation. This leads to processional motion i.e. spin torque oscillation.

STT induced magnetisation reversal as predicted by Slonczewski (1996) and Berger (1996) was demonstrated experimentally in a current-perpendicular-to-the-plane (CPP) GMR nanopillar by Albert et al. in 2000 where the STT device was produced by a combination of precise nano fabrication and controlled film growth. The critical current density for switching was reported to be $j_c \sim 10^7$ A/cm^2 which is at least an order of magnitude larger than the requirement for a Gbit MRAM device.

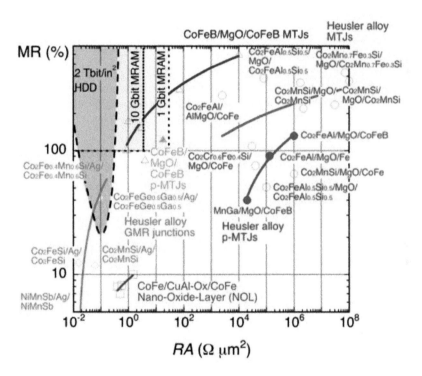

Fig. 8.4: Relationship between a MR ratio and RA of MTJs with CoFeB/MgO/CoFeB (blue triangles), nano-oxide layers (NOL, green squares) and Heusler alloys (red circles) with in-plane (open symbols) and perpendicular magnetic anisotropy (closed symbols) together with that of GMR junctions with Heusler alloys (orange rhombus). The target requirements for 2 Tbit/in^2 HDD read heads as well as 1 and 10 Gbit MRAM applications are shown as purple and yellow shaded regions, respectively (Hirohata *et al.*, 2018).

For switching at high frequency a further critical parameter has to be considered for both MRAM devices and high frequency read heads used in hard drives. It is known that for high frequency switching the product of the resistance (R) of the device times its area (A) is a critical parameter. In Fig. 8.4 the relationship between the magneto-resistance achieved and the resistance area (RA) product is shown for a range of different material systems some of which are in current use where others are proposed for further increases in the data density also shown in the figure.

Further significant decreases in the required current density have been achieved by the insertion of additional layers such as a Ru layer in a pseudo spin valve nanopillar with an additional ferromagnetic layer.

Many of these difficulties and challenges in MRAM architecture for rapid read and write capabilities with the latter being undertaken by STT, were achieved with the switch from GMR to TMR structures originally with an Al_2O_3 barrier which gave an improved TMR ratio of around 30% and j_c of $<10^7$ A/cm^2 (Fuchs *et al.*, 2004).

Following the ground breaking work of Butler *et al.* (2001) who showed theoretically that significantly greater TMR ratios could be achieved using epitaxial growth of a ferromagnet onto MgO as discussed in Section 5.4, the use of MgO was widely adopted for MRAM architectures. This change led to individual devices being developed with a TMR ratio of $\sim$100% and j_c $\sim$6×10^6 A/cm^2 with a pulsed current duration of 100 ms (Yuasa *et al.*, 2004). Such technology was used to produce a 4-kbit spin RAM by Sony in 2005 (Hosomi *et al.*, 2005). This device had elements with an RA product of 20 $\Omega\mu m^2$ and a TMR ratio of 160% with a minimum writing current of 20 μA. Accordingly 2nd generation MRAM development was accelerated significantly including a 2-Mbit MRAM chip with a cell size of 1.6×1.6 μm^2 with a TMR ratio of 100% and a write current of 200 μA by Hitachi and Tohoku University in 2007 (Kawahara *et al.*, 2007).

For MRAM operation the reference voltage V_{half} is normally set at the voltage given by

$$V_{half} = \frac{V_h - V_l}{2} \tag{8.5}$$

where V_h and V_l are the output voltages at the highest and lowest resistance. For Gbit MRAM, V_{half} needs to be of the order of 100 mV or less which can be achieved by either introducing a double tunnel junction with a ferromagnetic free middle layer or fabricating a textured tunnel junction. In STT-MRAM the supply voltage is 2.7-3.6 V which is almost the same as that for toggle MRAM. The corresponding stand by (read/write) current was reduced to 160 mA compared to 660 mA for the toggle MRAM system.

For a 1 Gbit MRAM the junction cell diameter should be <65 nm with a resistance area RA product <30 Ωm^2 and an MR ratio >100%. For a 10 Gbit MRAM the cell diameter has to be reduced to <20 nm with an $RA<3$-5 Ωm^2 and an MR ratio >100%. Here a low RA product is required to satisfy the need for impedance matching at high frequency and a low power consumption of <100 fJ/bit.

The standard MRAM architecture which has been employed commercially is to have each MRAM cell with a transistor attached with a large MR ratio of > 150% which is essential to maintain the signal-to-noise ratio allowing for a read-out signal voltage to be detected with only a small current applied. In-plane CoFeB/MgO/CoFeB MTJs have successfully satisfied the requirement for a 10 bit Gbit MRAM by achieving an RA as low as 0.9 $\Omega \mu m^2$ and a TMR ratio of slightly greater than 100% at room temperature as shown by the open triangles and the blue line in Fig. 8.4. Here the critical current density (j_c) can be controlled by the thermal stability for a pulsed current longer than 10 ns while it becomes adiabatic for <10 ns.

$$j_c = j_{c0} \left\{ 1 - \frac{kT}{K_u V} ln \left(\frac{\tau_p}{\tau_0} \right) \right\} (> 10 \ ns \ pulse) \tag{8.6}$$

$$j_c = j_{c0} \left\{ 1 + \left(\frac{\tau_1}{\tau_p} \right) ln \left(\frac{\tau_p}{\tau_0} \right) \right\} (< 10 \ ns \ pulse) \tag{8.7}$$

$$\dot{j}_{c0} = \frac{\alpha e M_s t \left\{ H_{ext} \pm (H_K + 2\pi M_s) \right\}}{\hbar \left\{ P/ (1 + P^2 cos\theta) \right\}} \tag{8.8}$$

In these formulae τ is the pulse width as the pulse decreases, τ_0 is the inverse of the attempt frequency (= 1 ns), $\tau_1 = \frac{1}{\alpha\gamma(H_K+2\pi M_s)}$ where γ is the gyromagnetic ratio, H_K the anisotropy field and M_s the saturation magnetisation. α is the Gilbert damping constant, e the electronic charge, t the thickness of the ferromagnet, H_{ext} the external field, $\hbar$ is Planck's constant divided by 2π and P is the polarisation of the ferromagnet. For further improvement in MRAM operation the following issues need to be resolved: firstly a reduction of the power consumption including j_c and a reduction of the cell size. For the reduction of the power consumption the thermal stability of the free layer has to be compromised. A synthetic antiferromagnet such as CoFeB/Ru/CoFeB has been used as the free layer in an MRAM cell by AIST which resulted in a five-fold increase in thermal stability with only an 80% increase in the critical current density for writing. Such a design could in principle achieve a 10-Gbit MRAM device. The reduction of the cell size can be achieved by adopting perpendicularly magnetised ferromagnetic layers in the MTJ as currently used in 3rd generation MRAM described in Section 8.3.

8.3 Third Generation MRAM with Perpendicular Orientation

For reasons described in Chapter 7 associated with perpendicular magnetic recording on disk, it would be preferable to have MRAM structures where the easy axes lie perpendicular to the plane of the film. Not only does this increase potential data density, i.e. the density of the cells per unit area, but also would allow for far greater scalability which would mean that each individual cell could in principle be made much smaller. These developments have required significant advances in magnetics technology itself but also in the complex tooling required to produce such devices and in particular the successful etching of the cells down to small dimensions without damaging the magnetic structure. Again interestingly, with reference back to Section 7.5, similar scaling of a TMR read head in a hard drive also required the availability of tooling that was capable of making devices on a nanometric scale. These developments have happened over the last decade or so i.e. from 2010 onwards. Again as discussed in Section 7.5 to induce a perpendicular anisotropy in any thin film requires the criterion for the anisotropy constant to be given as

$$K > 2\pi M_s^2 \qquad (8.9)$$

As discussed in Sections 7.3 and 7.4, such anisotropies can be achieved using alloys of materials with a large spin-orbit coupling such as Pt, Ir and Au. For reasons of cost, typically materials such CoPt, CoPd or multilayers of these materials have this required level of anisotropy. However as discussed in Section 7.6, the highest anisotropy of all which will give the highest thermal stability to a very small element is available from the alloy FePt in the L1$_0$-ordered phase. A selection of these alloys is shown in Fig. 8.5.

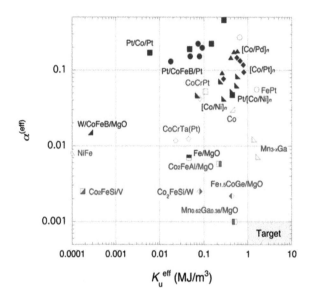

Fig. 8.5: Relationship between the magnetic anisotropy constant K_u^{eff} and the Gilbert damping constant α. Single films, multilayers with heavy metals and half-metallic Heusler alloy films are shown in green open, blue closed and red open symbols. Heusler alloys with MgO and heavy metals are also shown in half-closed symbols (Hirohata *et al.*, 2020).

While such layers can be used as the hard layer in an MTJ, FePt requires very high temperature annealing to be deposited in the L1$_0$ phase. Even with doping, temperatures of over 300°C are required to induce the necessary perpendicular anisotropy.

In a remarkable development, workers at the Spintec Laboratory in Grenoble and Tohoku University in Japan, found that if a common cubic ferromagnet such as Fe was grown on top of MgO the interface induced a perpendicular anisotropy in CoFeB (Manchon *et al.*, 2008). There is no simple physically intuitive way to describe how a material such as Fe which is cubic, and has a value of $K = 4.5 \times 10^5$ ergs/cc and easy axis along the cube edges, can have such a high anisotropy as per the requirement for in eqn 8.9. In order to understand and calculate how this happens, it has been found necessary to use extremely complex and sophisticated atomic based computer models of the structures and interactions that give rise to this effect. The most commonly used software is the well established Density Functional Theory (DFT). The best physical explanation in a pictorial sense relates back to the discussion in Section 2.7 where we are concerned with transition metal oxides and ferrimagnets. The spin ordering in those materials comes about because of the necessary bonding between the two p orbitals on the oxygen and the 3d orbitals on the transition metals as shown in Fig. 2.15. The nature of that bonding results in the common transition metal oxides ordering magnetically in an antiferromagnetic phase. A similar effect has been shown by computer modelling to occur when Fe and its alloys, importantly CoFeB are grown on an MgO surface. At the surface it has been shown that there is interfacial hybridisation between the electronic orbitals of the transition metal and those of the oxygen in the MgO layer. However this places a requirement on the growth process that the surface of the MgO layer be oxygen terminated because such an effect would not occur if the Fe was adjacent to Mg. The required epitaxial growth is shown in Fig. 8.6 which shows the required rotation of the Fe lattice with respect to the MgO (Ong *et al.*, 2015).

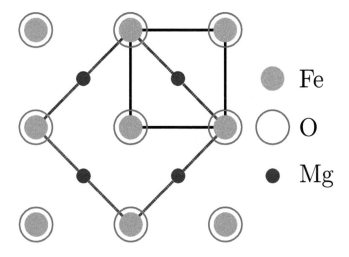

Fig. 8.6: Atomic configuration at the CoFe/MgO interface with O atoms sitting on top of Fe atoms (Ong *et al.*, 2015).

Advances in thin film growth techniques and materials processing have allowed this to be achieved. This effect is also highly convenient for STT-MRAM because MgO is also capable of forming an ideal tunnel barrier and $Fe_{60}Co_{40}$ has the highest value of M_s. It has been found that the anisotropy resulting from this interfacial hybridisation is of the order of 1.4 erg/cm^2 or 1.4 mJ/m^2. This is comparable to the perpendicular anisotropy that can be found in alloys such as CoPt. Again the layers are grown using the Anelva process whereby the as-deposited layers of MgO and in particular CoFeB are in an amorphous phase. The Fe-alloy can be crystallised at relatively modest temperatures and as it crystallises, because of the lattice match to Fe, it causes the MgO to crystallise. However in order for this to be the case and to induce the strong perpendicular interfacial anisotropy, the boron has to migrate out of the alloy and another layer known as a boron-getter is located at the interface of the alloy layer which attracts the boron towards it. This so-called getter layer is generally made of Ta, W or Mo (Dieny and Chshiev, 2017).

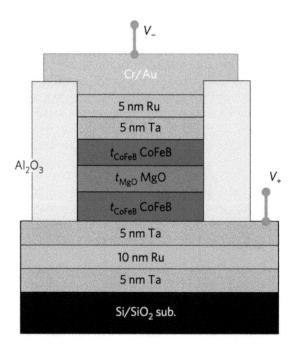

Fig. 8.7: Schematic of an MTJ device for MRAM applications. A perpendicular-anisotropy CoFeB–MgO magnetic tunnel junction, Ikeda *et al.*, Nature Materials, 9 721 (2010), Springer Nature, reproduced with permission from SNCSC.

The discovery and demonstration of STT-MRAM with perpendicular anisotropy resulting from the growth of Fe on MgO has caused a burst of commercialisation of this technology. The first perpendicular MRAM system was demonstrated by the group of Hideo Ohno at Tohoky University and achieved the requirement for a 1 Gbit system with an RA product of 18 $\Omega\mu m^2$ and a TMR of 124% at room temperature (Ikeda *et al.*, 2010). This system is shown in Fig. 8.7. Further development has allowed the demonstration of a system with potential to go to 10 Gbit MRAM device. This was undertaken using a synthetic antiferromagnet similar to that used in first generation toggle MRAM or a coupling of a hard ferromagnetic layer (Devolver *et al.*, 2019). These structures are shown in Fig. 8.1. These perpendicular MTJs were implemented in the first generation 256 Mbit MRAM produced by Everspin in 2019. Similarly, the

Samsung Corporation of Korea shipped a 256 Mbit system with a 28 nm element size for embedded memory evaluation in 2019.

These early commercial MRAM chips required further development due to the effect of the Gilbert damping factor. This factor can be high in the reference layer of an MTJ while other layers must exhibit low Gilbert damping for example, in the free layer, which is to be switched. This is required so as to minimise the switching current. These almost contradictory requirements are a complex matter because large perpendicular anisotropy and damping both originate from spin-orbit coupling as shown in Fig. 8.5. One possible solution is to use half metallic Heusler alloys because they have a low density of states at the Fermi level which is the dominant parameter controlling the damping constant. For example an MTJ system consisting of $Co_2FeAl(1.2)/MgO(1.8)/Fe(0.1)/CoFeB(1.3)$ where the thicknesses are in nanometres, was reported to show a TMR value of 132% and an RA of 1×10^6 $\Omega \mu m^2$ at room temperature (Wen *et al.*, 2014). A perpendicularly magnetised seed layer can also be used to induce out of plane magnetisation in a Heusler alloy thanks to the interfacial exchange interaction resulting from the growth of the Fe layer on top of the MgO which then orients the CoFeB layer to also have perpendicular alignment. There are a number of other Heusler alloy systems which have been reported to have perpendicular anisotropy as summarised in Fig. 8.4 (Rodmacq *et al.*, 2008). Significant research on these more complex layers including the insertion of a second MgO layer are ongoing at the time of writing. The addition of a second MgO layer in principle generates double the level of perpendicular anisotropy due to the interfacial effect. Other developments at the present time are examining the use of a synthetic antiferromagnetic layer as the free layer.

An alternative strategy to increase the perpendicular anisotropy is the use of shape anisotropy in the pillar (Watanabe *et al.*, 2018). This is perhaps the most obvious way to increase the anisotropy and thereby the thermal stability of any stored data. For this to be achieved it is necessary to make significantly smaller elements down to a diameter of about 12 nm or less. The growth of a ferromagnetic layer with a thickness of 15 nm then produces a significant enhancement to the overall anisotropy of each individual cell. For the axial ratio of 1.25 the shape anisotropy of K_s is 1.8×10^6 ergs/cc.

There are also a number of strategies being investigated to increase the overall TMR ratio. This depends on the spin dependent band structure of the quantum well formed at the interface between the ferromagnetic and tunnelling barriers. It has been found that the insertion of a layer of Cr at the interface between the MgO and the Fe can give rise to a symmetric band configuration (Greulet *et al.*, 2007). Alternatively a resonant tunnelling junction through the quantum well state has also been suggested by forming a double tunnelling barrier made of MgAlO. This double MTJ has demonstrated a TMR oscillation for a ferromagnetic layer of thickness up top 12 nm (Tao *et al.*, 2015).

Hence whilst we already have commercially viable STT-MRAM systems, because this is such a relatively new technology and to some extent revolutionary because of its reliance on a complex interface anisotropy, there are many potential strategies which can be attempted in order to improve specific characteristics such as thermal stability and the level of TMR generated. Hence it is likely that within the next 5 years, or possibly slightly longer, new materials and structures will be discovered which will further enhance the performance of MRAM devices. Now that the commercialisation of STT-MRAM has led to the generation of funding in the major electronic companies around the world these developments are likely to accelerate.

It has been demonstrated that an MRAM storage device with relatively high data density can be embedded on top of a CMOS chip (Dieny *et al.*, 2020). This possibility has the potential to completely revolutionise computer architecture and decrease costs because there is then no need to have interconnects between the various devices and the speed of operation of the MRAM cell is significantly faster than is possible with a silicon-based device. Of course information stored within an STT-MRAM system is permanent but switchable storage. That gives rise to the possibility of instant start-up of computers and hence the common practice of leaving laptops and desktop PCs switched on to avoid the start-up delay, will hopefully soon be ended. Given the number of devices in the world which are left permanently switched on this can make a significant impact towards the reduction of CO_2 emissions and hence global warming.

8.4   Other Devices Based on MRAM

As discussed above the concept of Magnetic Random Access Memory which has been an idea that has taken over 20 years to achieve practical implementation, also gives rise to some unique physics that can lead to other high technology engineering applications. At this time these are under active consideration because the necessary fabrication technology to produce viable devices already exists for disk magnetic storage. Here we provide a brief summary of effects that are known to occur in such devices but at this time are some way off being implemented commercially.

8.4.1   Spin-Torque Oscillator

Microwave oscillation has been demonstrated via the angular dependence of STT without the application of an external magnetic field (Boulle *et al.*, 2007) in an MRAM nanopillar consisting of $Ni_{0.81}Fe_{0.19}(4)/Cu(10)/Ni_{0.81}Fe_{0.19}(15)$ with thicknesses in nanometres show the frequency dependent STT. A broader frequency width response can be achieved by increasing the thickness of the upper layer of the permalloy so that the oscillating frequency range can increase from 300 Hz to 1.1 GHz. Similar effects have been observed in other structures based on CoFeB/MgO/CoFeB such that frequencies in the vicinity of 7 GHz or even higher have been reported (Kubota *et al.*, 2008). The frequency of these devices can be controlled via a bias voltage. To date the line width of the microwave oscillation in an MTJ is of the order of 100 MHz which is much wider than that required for device applications. Typically a Q-factor of 10,000 is required for a reliable device leading to a line width which should be <1 MHz. These devices typically have an output power of about 0.1 μW which is three orders of magnitude larger than a conventional GMR based oscillator. In theory these nanopillars can be operated almost up to the ferromagnetic resonance frequency of a few tens of GHz which is an improvement over the current silicon-based devices. However by patterning the ferromagnetic layer or the entire stack into a nanopillar or disk the Q-factor decreases significantly due to the presence of the edges of the device which have a certain level of roughness. In order to overcome these issues new designs are needed to be developed for a nanoscale spin-torque oscillator.

8.4.2 Neuromorphic Computation

Another intriguing possibility with a voltage tunable nanoresistance in an MTJ nanopillar is to use such a device for neuromorphic computing (Borders *et al.*, 2016). The concept is to utilise resistance changes in a current induced magnetisation reversal process and to control them by taking a minor loop of the free layer. In a manner similar to that in an MRAM cell, the data is stored as a magnetisation direction in one of the ferromagnetic layers in an MTJ which can be reversed by applying a current. Phase synchronisation of a vortex type spin-torque oscillator array has also been demonstrated (Tsunegi *et al.*, 2018). Up to eight spin-torque oscillators are used to demonstrate long-term stability of the phase difference to better than 1.6ms and the noise power spectral density of the phase difference of -80 dB/Hz at a 1 kHz offset frequency. Recently probabilistic magnetisation reversal has also been used for this potential application as a statistical computational device. An array of MTJs consisting of the standard CoFeB/MgO/CoFeB structure has been employed to demonstrate the factor derivation of integers of up to 945.

Figure 8.8 shows a schematic diagram and layouts of various spintronic memory cells. In Fig. 8.8(a) is a conventional spin transfer torque, (b) is a spin-orbit torque configuration discussed below and (c) is a Voltage Controlled Magnetic Anisotropy (VCMA) or Me MRAM (Liao *et al.*, 2020).

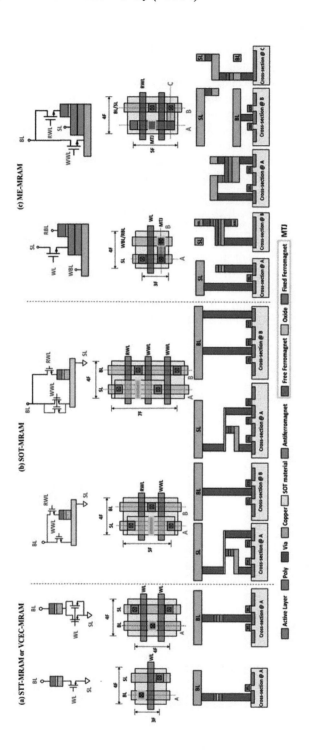

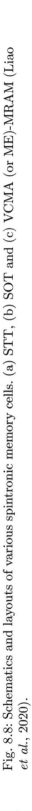

Fig. 8.8: Schematics and layouts of various spintronic memory cells. (a) STT, (b) SOT and (c) VCMA (or ME)-MRAM (Liao *et al.*, 2020).

8.4.3 Spin-Orbit Torque MRAM

Spin-Orbit Torque (SOT) results from a complex interaction between the spin of an electron and its orbit around a nucleus and the exchange interaction at an interface between typically a ferromagnetic metal and a noble metal such as gold or platinum. The mechanism of SOT arises from a relativistic consideration of the motion of an electron around a nucleus with atomic number Z. For simplicity we can consider a Bohr Model type of orbit such as that shown in Fig. 8.9. The electron is held in its orbit by the standard Coulomb force F_c. If we now take a view from the frame of reference of the electron, i.e. the electron is now stationary, then the charge on the electron *sees* the charge from the nucleus rotating. This constitutes a current loop which generates an effective magnetic field $\mathbf{B_{SO}}$ perpendicular to the plane of the orbit which is perpendicular to both the electron velocity $\mathbf{v}$ and the original electric field ($\mathbf{E}$) i.e.

$$\mathbf{B_{SO}} \simeq \frac{1}{c^2}\mathbf{v} \times \mathbf{E} \qquad (8.10)$$

where c is the speed of light. For complex quantum mechanical reasons $\mathbf{B_{SO}}$ is proportional to Z^4 and hence the effect is more pronounced for noble metals such as Pt, Pd and Au.

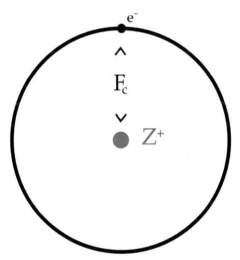

Fig. 8.9: Schematic diagram of an electron orbiting a nucleus with atomic number Z.

If we now consider a thin film of such a noble metal such as Pt shown in cross section in Fig. 8.10(a) then there will be a high density of conduction electrons moving at random within the lattice with spin values of $\pm\frac{1}{2}$. However if a current is now passed in the plane of the field then the current density $\mathbf{J}$ is given by

$$\mathbf{J} = |\sigma| \, \mathbf{E} \qquad (8.11)$$

where σ, the conductivity, is in the form of a matrix. **J** now gives rise to an average electron velocity in the plane of the film hence giving rise to a force perpendicular to the plane of the film. This will cause the conduction electrons to migrate to the upper and lower surfaces of the layer depending on the values of spin. This is analogous to the Stern-Gerlach experiment. Of course B_{SO} will be modulated due to the spatial distribution of the sites of the ions in the lattice. However, a pure spin current perpendicular to the plane is generated. This process leads to a spin accumulation of spin up and spin down electrons at either surface as shown schematically in Fig. 8.10(b). This spin current is capable of reversing a single domain element of a ferromagnetic element grown adjacent to the noble metal in a similar manner to spin transfer torque. Reversal of the element can proceed in either direction by simply reversing the direction of the current in the noble metal.

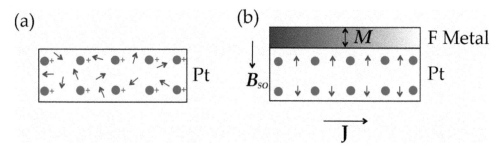

Fig. 8.10: Conduction electrons in a lattice with (a) $\mathbf{J} = 0$ and (b) $\mathbf{J}$ finite creating a pure spin current and spin accumulation.

Such spin-orbit torque switching can in principle produce devices in a similar manner to spin-transfer torque. However it is predicted theoretically and has been observed, that the switching can be much faster than is the case with spin-transfer torque and that the switching would involve significantly less energy than in a conventional MRAM device. The presence of this torque can also produce oscillatory behaviour in the ferromagnetic layer and again this has been observed in a tunnel junction consisting of the standard CoFeB/MgO/CoFeB structutre patterned on a Cu electrode (Liu *et al.*, 2012). Under a magnetic field application of -160 Oe an SOT torque oscillation has been observed in the GHz region.

Accordingly a new class of magnetic memory called Spin-Orbit Torque (SOT-MRAM) (Miron *et al.*, 2011) has been proposed to offer cache compatible high-speed and endurance at the cost of larger writing currents and the footprint of the device. In order to reduce the writing current and therefore the size of the selection transistor, charge spin conversion materials with low resistivity and large SOT efficiency must be developed. To be operated in stand-alone mode these devices require the application of a static magnetic field bias for which several embedded solutions have been proposed and some demonstrated on a large scale.

Because of the potential of this structure a number of industrial companies have already launched R&D projects in the area. In the context of this text however, this topic is not sufficiently near market for a detailed discussion to be included. However those interested in the topic are referred to a recent excellent review by Shao *et al.* published in 2021.

8.4.4   Voltage Controlled Magnetic Anisotropy

As long ago as the year 2000 the group of Hideo Ohno at Tohoku University in Japan reported on a magnetic phase transition induced by applying a bias voltage in a In-MnAl layer sandwiched within a field effect transistor (Ohno *et al.*, 2000). A positive bias on the gate of the device created an electric field which repels holes causing the Mn magnetic moment to rotate randomly in a manner similar to that in a paramagnet. A negative voltage generated an electric field which attracts the holes causing the Mn magnetic moment to align along the field direction hence exhibiting ferromagnetic order. These phenomena were only observed at a low temperature of 25 K with a bias voltage of ±125 V. This phenomenon is potentially a significant step towards the realisation of a fully electrically controlled spintronic device. If such a device could be created that would operate at room temperature, this would imply that the electric field could be used to lower the anisotropy of a ferromagnetic layer which would then switch using an effect similar to spin-transfer torque. Once the electric field is switched off, the resulting high anisotropy would be retained to create a storage device.

A similar control of magnetic behaviour has been demonstrated in an ultrathin Co film (Chiba *et al.*, 2011). The sample consisted of a 0.4 nm thick Co layer which is approximately one atom thick, in which a voltage of $+10$V induces ferromagnetism but -10 V induces non-magnetic behaviour. Similarly Ni nanoparticles dispersed between Ti and Pt layers exhibited a change between superparamagnetism and ferromagnetism with respect to the in-plane strain directions (Kim *et al.*, 2013). The voltage control is achieved via the electrical modulation of the spin-orbit interaction of the interfaces between a 3D transition metal ferromagnet and a dielectric layer.

Again this is another potential technique that could have application in an MRAM device. However at the present time such an application is believed to be at least half a decade away and hence further detail is not included at this time.

8.4.5   Thermally Assisted MRAM

The concept of a TA-MRAM system was proposed as long ago as 2007 by the Spintec Group in Grenoble, France (Prejbeanu *et al.*, 2007). This technology was based on the concept of an exchange bias MRAM system, albeit in-plane at the time. Basically a current pulse was used to selectively heat an exchange bias layer which would then reverse under a spin polarised current. Pulses as short as 500 ps were able to achieve writing. By controlling the blocking temperature at the antiferromagnetic $IrMn_3$ which pinned the ferromagnetic layer, operation was demonstrated with a direct current of 1.6 mA followed by the current pulse. The Joule heating in this way also may have potential to assist STT based magnetisation switching in a similar manner to heat assisted magnetic recording discussed in Section 7.6.

8.4.6 Technology Comparison

As discussed in the first part of this chapter, following many years of basic research and some level of technological development we now have a viable MRAM technology based on spin transfer torque and the use of perpendicular anisotropy. The importance of the concept of using an interfacial anisotropy to allow for this development of a viable technology cannot be understated. At the time of writing major companies such as Global Foundries in Singapore, Toshiba, Samsung and Intel are all in pre-production trials of such devices and notably several of these companies are looking at embedded memory where the MRAM storage is incorporated onto a CMOS chip.

However it is certainly the case that the concept of the solid-state magnetic memory based on remarkable discoveries and development, particularly in growth and etching techniques to allow for the required epitaxy and small scale cell size the whole concept of MRAM is probably in its infancy. In the latter part of this chapter we have reviewed several known phenomena which have potential to create 3rd, 4th and potentially 5th generation MRAM, all of which would have improved performance in terms of power consumption, storage density and in particular switching times.

Hence at the present time it is interesting to compare the already developed and potential different forms of MRAM with existing silicon-based technology and indeed with each other. Such a comparison table is shown in Table 8.3 but given the rate of development in this field the data provided there, which is up-to-date as far as we know, may well be out-of-date by the time this text is published.

Table 8.3 Comparison between MRAM and major memories (Hirohata and Takanashi, 2014; https://knowen.org/nodes/27127; https://www.extremetech.com/computing/163058-reram-the-new-memory-tech-that-will-eventually-replace-nand-flash-finally-comes-to-market; http://www.theilec.net/news/articleView.html?idxno=2428).

Technology	SRAM	eFlash	eDRAM	eSTT-MRAM	eSOT-MRAM	eMeRAM	eReRAM
Endurance	10^{16}	10^{15}	10^{15}	10^{15}	10^{15}	10^{15}	10^{5}
Read time (ns)	0.2~2	10	1~2	1~5	1~5	1~5	3~10
Write/erase time (ns)	0.2~2	$2.5\times10^{3}/2\times10^{6}$	1~2	5~10	<1	<1	$500/10^{5}$
Cell size (F^2)	92	40~100	40~100	40~50	50~70	20~30	15~30
Bit density (Mbit/cm^2)	64	$5\times10^{2} \sim 1\times10^{3}$	$5\times10^{2} \sim 10^{3}$	10^{3}	7.5×10^{2}	2×10^{3}	$(1.5\sim3)\times10^{3}$
Read energy (fJ/bit)	1~5	10^{6}	10^{2}	10~20	10~20	1~5	10^{3}
Write/erase energy (fJ/bit)	1	10^{6}	10^{3}	$(1\sim2)\times10^{2}$	<10	<5	$10^{3}/10^{6}$
Non-volatility	No	Yes	No	Yes	Yes	Yes	Yes
Standby power	Leakage	None	Refresh	None	None	None	None

9
Outlook for Future Developments

As we have seen in Part 1 of this text the theory and understanding of the physics underlying particulate and granular magnetism is well established and tested. Of course there remains the possibility of discovering new phenomena but these days such discoveries are increasingly rare with obvious exceptions such as the establishment of the interfacial anisotropy that can be generated between Fe and its alloys and MgO which has now been widely adapted for use in MRAM technology. However it is clear that such new phenomena can only be established, explained and exploited following the use of large scale computer simulations which can quantify and at least partly explain, these kinds of complex new effects. Given the range of the exchange interaction which leads to a necessary understanding of the complex phenomenon of particulate and granular magnetism at an atomic scale, it is unlikely that there will be any further significant developments in the basic understanding of this subject using analytical methods (see Fig. 7.26). Accordingly when looking to the future there is a clear division in the areas of application which can and probably will be developed, into those associated with information storage and those associated with magnetic particles dispersed in liquids. By far the largest by value and level of investment is of course in information storage.

9.1 Future Developments in Information Storage Technology

Future developments in information storage technology can themselves broadly be divided into two areas. The first of these is clearly hard drive technology which has probably been the most rapidly advancing area over the past 30 years. As indicated in Chapter 7 there is now a clear road map for a transition to Heat Assisted Magnetic Recording for very good reasons associated with data density and to some degree access time. At the time of writing it is in the public domain that Seagate Technology have actually shipped tens of thousands of HAMR drives to customers for testing and evaluation and aim to go into full production and distribution in the next couple of years. It seems therefore that the adoption of this technology is now inevitable and on a sound footing. That will probably increase the data density to something of the order of 4-6 Tbits/in^2. However beyond that it is difficult to see how hard drive technology could increase further in terms of data density but the amount of data stored in a single device can be increased by the simple expedient of putting more than one disk in a given box.

There has also been significant work in the academic field and to a limited extent in the industrial sphere, of looking at the possibility of using microwave energy to overcome the required very large anisotropy to achieve thermal stability at a very high density which in turn requires the use of very small bits. This technique is known as Microwave Assisted Magnetic Recording (MAMR) (Zhu *et al.*, 2008). The principle of MAMR is analogous to that for HAMR in that additional energy is applied during the write process so that very small, high anisotropy grains can be reversed but once reversed and following the removal of the microwave field, smaller bits can be written. It is commonly accepted that advances in technology require advances in basic science. However the adoption of technology is based almost exclusively on commercial or financial considerations. Given that HAMR is now established and already in the marketplace it would therefore be a requirement that the adoption of an alternative technology such as MAMR would have to show a significant financial incentive for one of the major data storage companies to adopt it for production. It seems unlikely that this will happen in the near future.

The second major advance in magnetic storage technology is that due to MRAM. At the time of writing a number of major companies are already in pre-production trials of this new technology for consumer use whereas previously its use was generally in specialist areas such as nuclear technology and military applications. This will constitute a new revolution in computer architecture particularly due to the possibility that MRAM storage can be grown directly on top of CMOS chips thereby removing the requirement for interconnects and generally simplifying the architecture of a computer itself. Of course given that the magnetic storage will be permanent, albeit switchable, this means that the start-up time for computers and the requirement to maintain some level of power to limit the time taken for loading the software, will make a significant impact for the end user. It is also likely that the use of MRAM technology will speed up significantly the rate at which a PC can achieve calculations and other operations.

The principle of what is likely to become the third generation of MRAM technology is now well established. As discussed in Chapter 8, it is based on the concept of STT-MRAM which involves all electrical switching and reading of the data and obviously no moving parts unlike a hard drive. The use of STT switching will reduce power consumption significantly and using similar techniques that were developed in the hard drive industry, it is likely that similar data densities will be achievable if not exceeding those achieved in the hard drive industry because of the need for servo tracks etc. which will not be needed. The most likely advances therefore in STT-MRAM will be associated with the ability to fabricate smaller and smaller elements rather than any significant change in the requirements for the granular materials involved in the fabrication of devices.

However in the last few years there has been a further significant advancement in the basic science of particulate and granular magnetism which may impact MRAM technology on perhaps a 3-5 year timescale. It had been established that a spin polarised current can reverse an element consisting of an antiferromagnetic material. In this sense, as discussed in Chapter 4 it is a reversal of the orientation of the spins within a small element rather than some transition involving the disorder of the antiferromagnetic spins themselves (Baltz *et al.*, 2018). This leads to the possibility of being able to generate information storage systems and perhaps other nano-electronic devices where due to the antiferromagnetic coupling, there will no longer be any demagnetising field which are present in ferromagnetic information storage. As shown in Chapter 4 as element sizes become ever smaller there can still be issues with an antiferromagnetic material in terms of thermal stability. It is the basis of the York Model of Exchange Bias and the York Protocols for the measurement of such systems, that small grains of an AF material can reorient the axis of the antiferromagnetic order due to thermal energy alone. Given that this is a completely new scientific phenomenon there will be some years of basic scientific research required to establish what is feasible. But as can be seen from the proceedings of recent conferences such as the ICM meeting in San Francisco in 2018 (*AIP Advances* vol. 8, Issue 10, October 2018), there is an enormous effort going into the fundamental science of AF spintronics particularly in academia.

Beyond these developments and even with blue sky thinking it is difficult to see how further advancements can be made in magnetic information storage. As discussed in Section 7.6.1 we are almost at the fundamental limits of what can be achieved given the availability of existing materials including quite complex and often exotic alloys. We therefore suspect that further advances in this area will rely on scaling to ever smaller sizes. This will depend critically on the anisotropy of the available materials as grain sizes will become so small such that, as has been the case throughout the history of magnetic information storage technology, the limitation will come from thermal energy.

9.2   Magnetic Fluid Systems

In the area of particles and grains of magnetic materials being incorporated into liquids, there has been relatively slow advances in the development of new materials and applications in this area when compared to dramatic advances made in information storage. For example in the field of magneto-rheological fluids discussed briefly in Section 6.7, these materials were first studied as long ago as 1948 (Rabinow, 1948). Applications of these materials did not really occur until the 1970s and 80s and then their primary use was in military vehicles as dampers for rough terrain vehicles. They were incorporated into certain luxury cars around the same time but it has only been in the last 15 years that their use has become more widespread.

Given that the requirement for the magnetic materials is to have high magnetisation and that currently iron or iron-alloy particles produced from the carbonyl iron decomposition discussed in Section 6.7 are used, there seems little possibility that any other economically viable material could be considered. Furthermore because the requirement is for rapid relaxation of the magnetisation of the particles it is necessary that they be of a size in the multidomain range typically $>1\mu$m. This would imply that the use of other materials are therefore precluded. Hence the only advances that can be made are in the increasingly widespread use of new devices primarily for damping purposes. Recently this has included their use in certain specialised bearings and even domestic devices such as washing machines for noise suppression.

Similarly sink-float separation of scrap materials in ferrofluids has been known about for over 30 years. However the cost of the fluids precluded widespread use of this effect. However as indicated in Section 6.4, there is now another attempt to undertake the separation of scrap materials and notably valuable metals, on a relatively large scale. This uses ferrofluids produced by grinding of large magnetite particles. It is interesting to note that the original ferrofluids prepared as long ago as in the 1960s used this technique before the co-precipitation method to produce magnetite nanoparticles was known.

The greatest area of potential advancement both in materials but also in applications, is in the sphere of biomedicine. The current and foreseeable requirements for the use of magnetic nanoparticles for cell separation or perhaps drug delivery is well satisfied by existing materials due to the restriction of in-vivo use to magnetite or maghemite and also due to the fact that these oxides have the highest moment possible. In particular the use of magnetic polymer spheres has already generated a very large worldwide industry. Current techniques and materials appear to satisfy all foreseeable demands.

The other areas where advances are possible are in direct medical technologies. The first of these that has recently received regulatory approval for in-vivo use is in the sphere of magnetic tracing. As discussed in Chapter 6, the use of magnetic tracing via the injection of suitably functionalised nanoparticles into the site of tumours and notably breast cancer, allows for the tracing of the spread of that disease to the lymphatic system via the use of a susceptometer. To our knowledge this is the first time such a technique has been developed and received approval and it may be that this breakthrough is the harbinger of further applications of this type in the sphere of therapeutic medicine.

Perhaps more disappointing has been the failure of magnetic hyperthermia to have developed more widely and quickly. There are often suggestions that magnetic hyperthermia could be a remedial treatment for cancer. That is probably not true because unless a cancerous growth is completely eliminated, the cancer is likely to recur. Also cancers such as that of the liver which are deeply embedded within the body, would require a complex and probably very large magnetic field to supply an adequate field at the site of the tumour.

However there are a number of other critically important areas where tumours occur close to the surface of the body where magnetic hyperthermia clearly has the potential to be of great efficacy. The first of these is frontal lobe brain tumours. These are generally non-malignant but do cause significant damage particularly to the optic nerve, and currently requires treatment using positron radiation. This relies on the positron annihilation effect when the positron meets an electron and generates gamma waves within the tumour that reduce its size. However it is almost always the case that some of the gamma rays escape and over a period of a few years this can result in hardening of the arteries of the brain. If magnetic hyperthermia could be applied to such tumours then they could be reduced in size preventing the harm caused for example to the eyesight but would not involve the use of ionising radiation. These tumours are very close to the surface of the skull and hence introduction of the magnetic nanoparticles would be very easy and also because of the proximity to the surface, generating a sufficient field of a few hundred Oe would also be relatively simple. Fortunately such tumours are generally very slow growing so perhaps a treatment once every few years would be all that would be required and could lead to a significant extension to life for those afflicted.

A much larger scale potential application relates to the problem in late middle aged and older men associated with the enlargement of the prostate gland. It is estimated that in Europe over half of men over 50 years old have some degree of enlargement to the prostate gland and fortunately only very few of these have a malignant enlargement. Existing treatments are mechanical in nature and hence relatively unpleasant and often surgery is required which can lead to some degree of incontinence. Again the prostate gland is very close to the surface and hence introducing the necessary magnetic field is relatively simple and likewise the introduction of the functionalised nanoparticles. We have described one potential technique that could make the treatment of both these problems far more effective because existing materials based on nanoparticles in water produce unpredictable and unreproducible heating due to the fact that the particles are free to move and can stir. We have shown that incorporation of specific tailored nanoparticles of a larger size into polymer spheres can now only reverse by magnetic hysteresis. This can generate sufficient heat to make a viable technology for near surface cell destruction.

For all these applications the requirement is simply to have a magnetic nanoparticle with appropriate properties which very often only involves a high magnetic moment and hence requires the use of magnetite. There is a need and almost certainly advances will be made in the magneto-chemistry of the growth of suitable materials such as that described in Section 6.5 which will allow further advances to occur without the need for more complex magnetic materials.

Appendix A

Demagnetising Factors for a Prolate and Oblate Spheroids

Table A.1 Demagnetising factors in the cgs system of units for a prolate ellipsoid (cigar-shaped) with semi–major axis c and semi-minor axis a ($N_c + 2N_a = 4\pi$).

c/a	$N_c/4\pi$	c/a	$N_c/4\pi$
1.0	0.333333	3.0	0.108709
1.1	0.308285	3.1	0.104410
1.2	0.286128	3.2	0.100376
1.3	0.266420	3.3	0.096584
1.4	0.248803	3.4	0.093015
1.5	0.232981	3.5	0.089651
1.6	0.218713	4.0	0.075407
1.7	0.205794	4.5	0.064450
1.8	0.194056	5.0	0.055821
1.9	0.183353	5.5	0.048890
2.0	0.173564	6.0	0.043230
2.1	0.164585	6.5	0.038541
2.2	0.156326	7.0	0.034609
2.3	0.148710	7.5	0.031275
2.4	0.141669	8.0	0.028421
2.5	0.135146	8.5	0.025958
2.6	0.129090	9.0	0.023816
2.7	0.123455	9.5	0.021939
2.8	0.118203	10.0	0.020286
2.9	0.113298	20.0	0.006749

Table A.2 Demagnetising factors in the cgs system of units for an oblate ellipsoid (squashed spheroid), i.e. $a > c$ and $N_a + 2N_c = 4\pi$.

c/a	$N_a/4\pi$	c/a	$N_a/4\pi$
1.0	0.333333	2.4	0.577227
1.1	0.359073	2.5	0.0.588154
1.2	0.383059	2.6	0.598539
1.3	0.405437	2.7	0.608422
1.4	0.426344	2.8	0.617837
1.5	0.445906	2.9	0.626817
1.6	0.464237	3.0	0.635389
1.7	0.481442	3.1	0.643581
1.8	0.497615	3.2	0.651417
1.9	0.512843	3.3	0.658920
2.0	0.527200	3.4	0.666110
2.1	0.540758	3.5	0.673006
2.2	0.553578	4.0	0.703641
2.3	0.565717	4.5	0.729061

References

Aharoni A. (1985). *J. Appl. Phys.*, **57**, 4702.

Åkerman J., De Herrera M., Durlam M., Engel B., Janesky J., Mancoff F., Slaughter J., Tehrani S. (2004). *Magnetic tunnel junction based magnetoresistive random access memory M. Johnson (Ed.), Magnetoelectronics (Elsevier, Amsterdam, Netherlands) p. 231-272.*

Albert F. J., Katine J., Buhrmn R. A., Ralph D. C. (2000). *Appl. Phys. Lett.*, **77**, 3809.

Aley N. P., Bonet C., Lafferty B., O'Grady K. (2009). *IEEE Trans. Magn.*, **45**, 3858.

Aley N. P., O'Grady K. (2011). *J. Appl. Phys.*, **109**, 07D719.

Aley N. P., Vallejo-Fernandez G., Kroeger R., Lafferty B., Agnew J., Lu Y., O'Grady K. (2008). *IEEE Trans. Magn.*, **44**, 2820.

Ali M., Marrows C. H., Hickey B. J. (2008). *Phys. Rev. B*, **77**, 134401.

Almeida T. P., Palomino A., Lequeux S., Boureau V., Fruchart O., Prejbeanu I. L., Dieny B., Cooper D. (2022). *Appl. Phys. Lett. Mater.*, **10**, 061104.

Baibich M. N., Broto J. M., Fert A., Nguyen Van Dau F., Petroff F., Etienne P., Creuzet G., Friederich A., Chazelas J. (1988). *Phys. Rev. Lett.*, **61**, 2472.

Baltz V., Manchon A., Tsoi M., Moriyama T., Ono T., Tserkovnyak Y. (2018). *Rev. Modern Phys.*, **90**, 015005.

Baltz V., Sort J., Landis S., Rodmacq B., Dieny B. (2005). *Phys. Rev. Lett.*, **94**, 117201.

Barkhausen H. (1919). *Phys. Z.*, **20**, 401.

Bean, C. P., Livingston J. D. (1959). *J. Appl. Phys. Suppl.*, **30**, 120S.

Berger L. (1996). *Phys. Rev. B*, **54**, 9353.

Berkowitz A. E., Greiner J. H. (1965). *J. Appl. Phys.*, **36**, 3330.

Berkowitz A., Takano K. (1999). *J. Magn. Magn. Mater.*, **200**, 552.

Berry C. C. (2009). *J. Phys. D: Appl. Phys.*, **42**, 224003.

Berry C. C., Curtis A. S. G. (2003). *J. Phys. D: Appl. Phys.*, **36**, R198.

Bethe H. (1933). *Handbuch der Physik. In: Geiger, H. and Scheel, K., Eds., Vol. 24, Part 1, Springer, Berlin..*

Binasch G., Grünberg P., Saurenbach F., Zinn W. (1989). *Phys. Rev. B*, **39**, 4828(R).

Bogardus E., Scranton R., Thompson D. (1975). *IEEE Trans. Magn.*, **11**, 1364.

Bolyachkina A, Sepehri-Amin H., Suzuki I., Tajiri H., Takahashi Y. K., Srinivasan K., Ho H., Yuan H., Seki T., Ajan A., Hono K. (2022). *Acta Materialia*, **227**, 1417744.

Borders W. A., Akima H., Fukami S., Moriya S., Kurihara S., Horio Y., Sato S., Ohno H. (2016). *Appl. Phys. Exp.*, **10**, 013007.

Boulle O., Cros V., Grollier J., Pereira L. G., Deranlot C., Petroff F., Faini G., Barnaś J., Fert A. (2007). *Nat. Phys.*, **3**, 492.

Butler W. H., Zhang X.-G., Schulthess T. C., MacLaren J. M. (2001). *Phys. Rev. B*, **63**, 054416.

Carpenter R., Cramp N. C., O'Grady K. (2012). *IEEE Trans. Magn.*, **48**, 4351.

Chantrell R. W., O'Grady K. (1992). *North-Holland Delta Series*, 103-113.

Chantrell R. W., Popplewell J., Charles, S. W. (1978). *IEEE Trans. Magn.*, **14**, 975.

Charles, S. W., Chandrasekhar R., O'Grady K., Walker M. (1988). *J. Appl. Phys.*, **64**, 5840.

Chiba D., Fukami S., Shimamura K., Ishiwata N., Kobayashi K., Ono T. (2011). *Nat. Mater.*, **10**, 853.

Chikazumi O. (1997). *Physics of Ferromagnetism (2nd edn.) Oxford Science Publications*.

Chureemart J., Lari L., Nolan T. P., O'Grady K. (2013). *J. Appl. Phys.*, **114**, 083907.

Clarke D. M., Vallejo-Fernandez. (2022). *J. Magn. Magn. Mater.*, **552**, 169249.

Clarke D.M., Marquina C., Lloyd D.C., Vallejo-Fernandez G.(2022). *J. Magn. Magn. Mater.*, **559**, 169543.

Cullity B. D. (1972). *Introduction to Magnetic Materials (1st edn.) Addison-Wesley*.

Cullity B. D., Graham C. D. (2008). *Introduction to Magnetic Materials (2nd edn.) IEEE Press*.

Daughton J. M. (1997). *J. Appl. Phys.*, **81**, 3758.

de Witte A. M., el-Hilo M., O'Grady K., Chantrell R. W. (1993). *J. Magn. Magn. Mater.*, **120**, 184.

de Witte A. M., O'Grady K., Coverdale G. N., Chantrell R. W. (1990). *J. Magn. Magn. Mater.*, **88**, 183.

de Witte A. M., O'Grady K. (1990). *IEEE Trans. Magn.*, **26**, 1810.

DeBrosse J., Gogl D., Bette A., Hoenigschmid H., Robertazzi R., Arndt C., Braun D., Casarotto D., Havreluk R., Lammers S., Obermaier W., Reohr W. R., Viehmann H., Gallagher W. J., Müller G. (2004). *IEEE J. Solid-State Circ.*, **39**, 678.

Devolder T., Carpenter R., Rao S., Kim W., Couet S., Swerts J., Kar G. S. (2019). *J. Phys. D: Appl. Phys.*, **52**, 274001.

Dickson D. P. E., Reid N. M. K., Hunt C., Williams H. D., el-Hilo M., O'Grady K. (1993). *J. Magn. Magn. Mater.*, **125**, 345.

Dieny B., Chshiev M. (2017). *Rev. Mod. Phys.*, **89**, 025008.

Dieny B., Prejbeanu L. J., Garello K., Gambardella P., Freitas P., Lehndorff R., Raberg W., Ebels U., Demokritov S. O., Akerman J., Deac A., Pirro P., Adelmann C., Anane A., Chumak A. V., Hirohata A., Mangin S., Valenzuela S. O., Cengiz Onbaşlı M. , d'Aquino M., Prenat G., Finocchio G., Lopez-Diaz L., Chantrell R. W. (2020). *Nature Electronics*, **3**, 446.

Dieny B., Speriosu V. S., Parkin S. S. P., Gurney B. A., Wilhoit D. R., Mauri D. (1991). *Phys. Rev. B*, **43**, 1297(R).

Doyle W. D., He L., Flanders P. J. (1993). *IEEE Trans. Magn.*, **29**, 3634.

Drayton A., Zehner J., Timmis J., Patel V., Vallejo-Fernandez G., O'Grady K. (2017). *J. Phys. D: Appl. Phys.*, **50**, 495003.

Dutson J. D., Hürrich C., Vallejo-Fernandez G., Fernandez-Outon L. E., Yi G., Mao S., Chantrell R. W., O'Grady K. (2007). *J. Phys. D: Appl. Phys.*, **40**, 1293.

Dutson J. D., O'Grady K., Lu B., Kubota Y., Platt C. L. (2003). *IEEE Trans. Magn.*, **39**, 2344.

el-Hilo M., O'Grady K., Popplewell J. (1991). *J. Appl. Phys.*, **69**, 5133.

el-Hilo M., O'Grady K., Chantrell R. W. (1992a). *J. Magn. Magn. Mater.*, **109**, L164.

el-Hilo M., de Witte A. M., O'Grady K., Chantrell R. W. (1992b). *J. Magn. Magn. Mater.*, **117**, L307.

el-Hilo M., O'Grady K., Chantrell R. W. (1992c). *J. Magn. Magn. Mater.*, **114**, 295.

el-Hilo M., O'Grady K., Chantrell R. W. (1992d). *J. Magn. Magn. Mater.*, **114**, 307.

el-Hilo M., O'Grady K., Chantrell R. W. (1993). *J. Appl. Phys.*, **73**, 6653.

el-Hilo M., O'Grady K., Chantrell R. W. (2002). *J. Magn. Magn. Mater.*, **248**, 360.

Engel B. N., Åkerman J., Butcher B., Dave R. W., De Herrera M., Durlam M., Grynkewich G., Janesky J., Pietambaram S. V., Rizzo N. D., Slaughter J. M., Smith K., Sun J. J., Tehrani S.(2005). *IEEE Trans. Magn.*, **41**, 132.

Englert E. (1932). *Ann. Phys., Lpz.* , **14**, 589.

Estrin Y., McCormick P. G., Street R. (1989). *J. Phys.: Condens. Matter*, **1**, 4845.

Ewing J. A. (1891). *Proc. Roy. Soc. Lond.*, **48**, 342.

Faraday M. (1839). *Experimental Researches in Magnetism, J. M. Dent and Sons*,

Farrow R. F. C., Marks R. F., Gider S., Marley A. C., Parkin S. S. P., Mauri D. (1997). *J. Appl. Phys.*, **81**, 4986.

Fecioru-Morariu M., Ali S. R., Papusoi C., Sperlich M., Guntherodt G. (2007). *Phys. Rev. Lett.*, **99**, 097206.

Fernandez-Outon, L. E., Carey M. J., O'Grady K. (2004). *J. Appl. Phys.*, **95**, 6852.

Fernandez-Outon, L. E., Vallejo-Fernandez G., Manzoor S., Hillebrands B., O'Grady K. (2008). *J. Appl. Phys.*, **104**, 093907.

Fernandez-Outon, L. E., Vallejo-Fernandez G., O'Grady K. (2006). *J. Magn. Magn. Mater.*, **303**, 296.

Ferreira H. A., Graham D. L., Freitas P. P., Cabral J. M. S. (2003). *J. Appl. Phys.*, **93**, 7281.

Fert A., Campbell I. A. (1968). *Phys. Rev. Lett.,*, **21**, 1190.

Fert A., Campbell I. A. (1976). *J. Phys. F: Metal Phys.,*, **6**, 849.

Ford P. J. (1982). *Contemp. Phys.*, **23**, 141.

Frenkel J., Doefman J. (1930). *Nature*, **126**, 274.

Fuchs G. D., Emley N. C., Krivorotov I. N., Braganca P. M., Ryan E. M., Kiselev S. I., Sankey J. C., Ralph D. C., Buhrman R. A., Katine J. A. (2004). *Appl. Phys. Lett.*, **85**, 1205.

Fulcomer E., Charap S. H. (1972a). *J. Appl. Phys.*, **43**, 4184.

Fulcomer E., Charap S. H. (1972b). *J. Appl. Phys.*, **43**, 4190.

Futamoto M., Inaba N., Hirayama Y., Ito K., Honda Y. (1998). *MRS Online Conf. Proc.*, **517**, 243.

Gaunt, P. (1986). *J. Appl. Phys.*, **59**, 4129.

Gilbert W. (1600). *De Magnete, Magneticisque Corporibus, et de Magno Magnete Tellure Peter Short, London (1st edition, in Latin).*

Gilchrist R. K., Medal R., Shorey W. D., Hanselman R. C., Parrott J. C., Taylor C. B. (1957). *Ann. Surg.*, **146**, 596.

Gittleman J. I., Abeles B., Bozowski S. (1974). *Phys. Rev.*, **9**, 3891.

Gittleman J. I., Goldstein Y., Bozowski S. (1972). *Phys. Rev. B*, **5**, 3609.

Gneveckow U., Jordan A., Scholz R., V. Brüss, Waldöfner N., Ricke J., Feussner A., Hildebrandt B., Rau B., Wust P. (2004). *Med. Phys.*, **31**, 1444.

Goldfarb R. (2017). *IEEE Magn. Lett.*, **8**, 1110003.

Goldfarb R. (2018). *IEEE Magn. Lett.*, **9**, 1205905.

Gompertz J., Carpenter R., Hassan S., Ormston M., O'Grady K. (2022). *IEEE Trans. Magn.*, **58**, 4800105.

Granley G. B., Daughton J. M., Pohm A. V., Comstock C. S. (1991). *IEEE Trans. Magn.*, **27, 5517**.

Granqvist C. G., Buhrman R. A. (1976). *J. Appl. Phys.*, **47**, 2200.

Greulet F., Tiusan C., Montaigne F., Hehn D., Halley D., Bengone O., Bowen M., Weber W. (2007). *Phys. Rev. Lett.*, **99**, 187202.

Hayashi M., Thomas L., Moriya R., Rettner C., Parkin S. S. P. (2008). *Science*, **320**, 209.

Heisenberg W. (1928). *Zeitschrift für Physik.*, **49**, 619.

Henkel O. (1964). *Phys. Stat. Sol.*, **7**, 919.

Hirohata A., Frost W., Samiepour M., Kim J.-Y. (2018). *Materials*, **11**, 105.

Hirohata A., Yamada K., Nakatani Y., Prejbeanu L., Dieny B., Pirro P., Hillebrand B. (2020). *J. Magn. Magn. Mater.*, **509**, 166711.

Hisano S., Saito K. (1998). *J. Magn. Magn. Mater.*, **190**, 371.

Honda K., Kaya S. (1926). *Sci. Rep. Tohoku*, **15**, 721.

Hosomi M., Yamagishi H., Yamamoto T., Bessho K., Higo Y., Yamane K, Yamada H., Shoji M., Hachino H., Fukumoto C., Nagao H., Kano H. (2005). *Int. Electron Devices Meeting (IEDM) Tech. Dig.*, 459.

Hubert A., Schäfer R. (1998). *Magnetic Domains: The analysis of magnetic microstructures, Springer Berlin, Heidelberg.*

Hussain R., Kaeswurm B., O'Grady K. (2011). *J. Appl. Phys.* **109**, 07E533.

Ikeda S., Miura K., Yamamoto H., Mizunuma K., Gan H. D., Endo M., Kanai S., Hayakawa J., Matsukura F., Ohno H. (2010). *Nat. Mater.*, **9**, 721.

Inomata K., Koike N., Nozaki T., Abe S., Tezuka N. (2003). *Appl. Phys. Lett.*, **82**, 2267.

Inoue K., Shima H., Fujita A., Ishida K., Oikawa K., Fukamichi K. (2006). *Appl. Phys. Lett.*, **88**, 102503.

Iwasaki S., Nakamura Y., Ouchi K. (1979). *IEEE Trans. Magn.*, **15**, 1456.

Jacobs I. S., Bean C. P. (1955). *Phys. Rev.*, **100**, 1060.

Jalli J., Hong Y.-K., Bae S., Abo G. S., Lee J.-J., Sur J.-C., S. H. Gee, Kim S.-G., Erwin S. C., Moitra A. (2009). *IEEE Trans. Magn.*, **45**, 3590.

Jiles D. C. (1992). *Introduction to Magnetism and Magnetic Materials (1st edn.) Chapman and Hall, London.*

Joffe I., Heuberger R. (1974). *Phil. Mag.*, **314**, 1051.

Johannsen M., Gneveckow U., Thiesen B., Taymoorian K., Cho C. H., Waldofner N., Scholz R., Jordan A., Loening S. A., Wust P. (2007). *European Urology*, **52**, 1653.

Julliere M. (1975). *Phys. Lett. A*, **54**, 225.

Katine J. A., Albert F. J., Buhrman R. A., Myers E. B., Ralph D. C. (2000). *Phys. Rev. Lett.*, **84**, 3149.

Kasuya T. (1956). *Prog. Theor. Phys.* , **16**, 45.

Kawahara T., Takemura R., Miura K., Hayakawa J., Ikeda S., Lee Y., Sasaki R., Goto Y., Ito K., Meguro T., Matsukura F., Takahashi H., Matsuoka H., Ohno H. (2007). *IEEE International Solid-State Circuits Conference (ISSCC) Tech. Dig.* , 480.

Kelly P. E., O'Grady K., Mayo P. I., Chantrell R. W. (1989). *IEEE Trans. Magn.*, **25**, 3881.

Khalafalla S. E., Reimens G. W. (1980). *IEEE Trans. Magn.*, **16**, 178.

Kim Y. K., Park G. H., Lee S. R., Min S. H., Won J. Y. (2003). *J. Appl. Phys.*, **93**, 7924.

Kim H. K. D., Schelhas L. T., Keller S., Hockel J. L., Tolbert S. H., Carman G. P. (2013). *J. Nano Lett.*, **13**, 884.

Kittel C. (1946). *Phys. Rev.*, **70**, 965.

Kittel C., Galt J. K. (1956). *Solid State Phys.*, **3**, 437.

Kittel C., Galt J. K.,Campbell W. E. (1950). *Phys. Rev.*, **77**, 725.

Kneller E. (1964). *Proc. Int. Conf. Magn. (Physical Society of London 1975.*,174.

Koester E. (1984). *IEEE Trans. Magn.*, **20**, 81.

Koon N. C. (1997). *Phys. Rev. Lett.*, **78**, 4865.

Kubota H., Fukushima A., Yakushiji K., Nagahama T., Yuasa S., Ando K., Maehara H., Nagamine Y., Tsunekawa K., Djayaprawira D. D., Watanabe N., Suzuki Y. (2008). *Nat. Phys.*, **4**, 37.

Kurenkov A., Fukami S., Ohno H. (2020). *J. Appl. Phys.*, **128**, 010902.

Lagae L., Wirix-Speetjens R., Liu C.-X., Laureyn W., Borghs G., Harvey S., Galvin P., Ferreira H. A., Graham D. L., Freitas P. P., Clarke L. A., Amaral M. D. (2005). *IEE Proc.-Circuits Devices Syst.*, **152**, 393.

Lewis V. G., Mayo P. I., O'Grady K. (1993). *J. Appl. Phys.*, **73**, 6656.

Liao Y.-C., Pan C., Naeemi A. (2020). *IEEE J. Explor. Solid-State Comput. Devices Circuits*, **6**, 9.

Liu L., Pai C.-F., Ralph D. C., Buhrman R. A. (2012). *Phys. Rev. Lett.*, **109**, 186602.

Luborsky F. E. (1961). *J. Appl. Phys.*, **32**, S171.

Maekawa S., Gafvert U. (1982). *IEEE Trans. Magn.*, **18**, 707.

Mahmoudi K., Bouras A., Bozec D., Ivkov R., Hadjipanayis C. (2018). *Int. J. Hyperthermia*, **34**, 1316.

Maier-Hauff K, Rothe R., Scholz R., Gneveckow U., Wust P., Thiesen B., Feussner A., von Deimling A., Waldoefner N., Felix R., Jordan A. (2007). *J. Neurooncol.*, **81**, 53.

Malozemoff A. P.(1987). *Phys. Rev. B*, **35**, 3679.

Manchon A., Ducruet C., Lombard L., Auffret S., Rodmacq B., Dieny B., Pizzini S., Vogel J., Uhlíř V., Hochstrasser M., Panaccione G. (2008). *J. Appl. Phys.*, **104**, 043914.

Marchon B., Pitchford T., Hsia Y.-T., Gangopadhyay S. (2013). *Adv. Trib.*, 521086.

Mauri D., Kay E., Scholl D., Howard J. K.(1987). *J. Appl. Phys.*, **62**, 2929.

Mayo P. I., O'Grady K., Chantrell R. W., Cambridge J. A., Sanders I. L., Yogi T., Howard J. K. (1991). *J. Magn. Magn. Mater.*, **95**, 109.

McGhie A., Marquina C., O'Grady K., Vallejo-Fernandez G. (2017). *J. Phys. D: Appl. Phys.*, **50**, 455003.

Meiklejohn W. H. (1962). *J. Appl. Phys.*, **33**, 1328.

Meiklejohn W. H., Bean C. P. (1956). *Phys. Rev.*, **102**, 1413.

Menear S., Bradbury A., Chantrell R. W. (1984). *J. Magn. Magn. Mater.*, **43**, 166.

Men'shikov A. Z., Takzei G. A., Dorofeev Y. A., Kazantsev V. A., Kostyshin A. K., Sych I. I. (1985). *Sov. Phys. JETP*, **62**, 734.

Michelson A. A., Morley E. W. (1887). *American J. Sci.*, **34**, 333.

Miron I. M., Garello K., Gaudin G., Zermatten P.-J., Costache M. V., Auffret S., Bandiera S., Rodmacq B., Schuhl A., Gambardella P. (2011). *Nature*, **476**, 189.

Molday R. S., MacKenzie D. (1982). *J. Immunol. Methods*, **52**, 353.

Moodera J. S., Kinder L. R., Wong T. M., Meservey R. (1995). *Phys. Rev. Lett.*, **74**, 3273.

Morrish A. H., Yu S. P. (1956). *Phys. Rev.*, **102**, 670.

Mott N. F. (1936). *Proc. Roy. Soc. A*, **153**, 699.

Nakamura S., Saito Y., Morise H.(2006). *Toshiba Rev.*, **61**, 40.

Néel, L. (1948). *Ann. Phys.*, **3**, 137.

Néel, L. (1949a). *Comp. Rend. Acad. Sci.*, **228**, 664.

Néel, L. (1949b). *Ann. Geophys.*, **5**, 99.

Néel, L. (1959). *J. de Phys.*, **20**, 215.

Néel, L. (1967). *Ann. Phys. (Paris)*, **2**, 61.

Nolan T. P., Valcu B. F., Richter H. J. (2011). *IEEE Trans. Magn.*, **47**, 63.

Nowak U., Usadel K. D., Keller J., Miltén“yi P., Beschoten B., Guntherodt G. (2002). *Phys. Rev. B*, **66**, 014430.

O'Grady K. (2022). *UK Patent* , Number 2211256.9, 2nd August .

O'Grady K., Bradbury A. (1983). *J. Magn. Magn. Mater.*, **39**, 91.

O'Grady K., Bradbury A., Charles S. W., Menear S., Popplewell J., Chantrell R. W. (1983b). *J. Magn. Magn. Mater.*, **31-34**, 958.

O'Grady K., Chantrell R. W. (1992). *In Magnetic Properties of Fine Particles, Eds. Dormann and Fiorani, North-Holland Delta Series*, , 93-102.

O'Grady K., Chantrell R. W., Popplewell J., Charles S. W. (1981). *IEEE Trans. Magn.*, **17**, 2943.

O'Grady K., Dova P., Laidler H. (1998). *Mat. Res. Soc. Symp. Proc.*, **417**, 231.

O'Grady K., el-Hilo M., Chantrell R. W. (1993). *IEEE Trans. Magn.*, **29**, 2608.

O'Grady K., Greaves S. (1995). *IEEE Trans. Magn.*, **31**, 2794.

O'Grady K., Popplewell J., Charles S. W. (1983a). *J. Magn. Magn. Mater.*, **39**, 56.

O'Grady K., Vallejo-Fernandez G., Fernandez-Outon, L. E. (2010). *J. Magn. Magn. Mater.*, **322**, 883.

O'Grady K., Walmsley N. S., Wood C. F., Chantrell R. W. (1998). *IEEE Trans. Magn.*, **34**, 1579.

O'Handley R. C. (1999). *Modern Magnetic Materials: Principles and Applications (1st edn.) Wiley*.

Oersted H. C. (1820). *Annals of Philosophy*, **16**, 273.

Ohno H., Chiba D., Matsukura F., Omiya T., Abe E., Dietl T., Ohno Y., Ohtani K. (2000). *Nature,* **408**, 944.

Oldham K., Felix S., Conway R., Horowitz R. (2008). *Proceedings of the American Control Conference*, art. no. 4587187, 4400.

Ong P. V., Kioussis N., P. K. Amiri, Wang K. L., Carman G. P. (2015). *J. Appl. Phys.*, **117**, 17B518.

Page C. H. (1970). *American J. Phys.*, **38**, 421.

Pankhurst Q. A., Connolly J., Jones S. K., Dobson J. (2003). *J. Phys. D: Appl. Phys.*, **36**, R167.

Pankhurst Q. A.,Thanh N. K. T.,, Jones S. K., Dobson J. (2009). *J. Phys. D: Appl. Phys.*, **42**, 224001.

Parkin S. S. P. (1991). *Phys. Rev. Lett.*, **67**, 3598.

Parkin S. S. P., Hayashi M, Thomas L. (2008). *Science*, **320**, 190.

Parkin S. S. P., Kaiser C., Panchula A., Rice P. M., Hughes B., Samant M., Yang S.-H. (2004). *Nat. Mater.*, **3**, 862.

Parkin S. S. P., Li Z. G., Smith J. (1991). *Appl. Phys. Lett.*, **58**, 2710.

Popplewell J., Davies P. C., Llewellyn J. P., O'Grady K. (1986). *J. Magn. Magn. Mater.*, **54-57**, 761.

Poulsen V. (1899). *US Patent* , Number 661619.

Prejbeanu I. L., Kerekes M., Sousa R. C., Sibuet H., Redon O., Dieny B., Nozières J. P. (2007). *J. Phys.: Condens. Matter*, **19**, 165218.

Rabinow J. (1948). *AIEE Trans.*, **67**, 1308.

Roca A. G., Costo R., Rebolledo A. F., Veintemillas-Verdaguer S., Tartaj P., Gonzalez-Carreno T., Morales M. P., Serna C. J. (2009). *J. Phys. D: Appl. Phys.*, **42**, 224002.

Rodmacq B., Auffret S., Dieny B., Nistor L. E. (2008). *US Patent* , Number 8513944B2.

Rosensweig, R. E. (1985). *Ferrohydrodynamics, Cambridge University Press (Cambridge Monographs on Mechanics and Applied Mathematics)*.

Rosensweig, R. E. (2002). *J. Magn. Magn. Mater.*, **252**, 370.

Rubia-Rodríguez I., Santana-Otero A., Spassov S., Tombácz E., Johansson C., de La Presa P., Teran F. J., del Puerto Morales M., Veintemillas-Verdaguer S., Thanh n. T. K., Besenhard M. O., Wilhelm C., Gazeau F., Harmer Q., Mayes E., Manshian B. B., Soenen S. J., Gu Y., Millán A., Efthimiadou E. K., Gaudet J., Goodwill P., Mansfield J., Steinhoff U., Wells J., Wiekhorstm F., Ortega D. (2021). *Materials*, **14**, 706.

Ruderman M., Kittel C. (1954). *Phys. Rev.*, **96**, 99.

Saha J., Victora R. H. (2006). *Phys. Rev. B*, **73**, 104433.

Saito Y. (2021). *Magnetic random access memory in A. Yamaguchi, A. Hirohata and B. Stadler (Eds.), Nanomagnetic Materials: Fabrication, Characterization and Applications (Elsevier, Amsterdam, Netherlands) p. 486-501*.

Sasaki I., Nakatani R., Ishimoto K., Endo Y., Shiratsuchi Y., Kawamura Y., Yamamoto M. (2007). *J. Magn. Magn. Mater.*, **310**, 2677.

Schabes W. E., Bertram H. N. (1988). *J. Appl. Phys.*, **64**, 5832.

Schwee L. J. (1972). *IEEE Trans. Magn.*, **8**, 405.

Scholten G., Usadel K. D., Nowak U. (2005). *Phys. Rev. B*, **71**, 064413.

Schull C. G., Strauser W. A., Wolan E. O. (1951). *Phys. Rev.*, **83**, 333.

Schulthess T.C., Butler W.H. (1998). *Phys. Rev. Lett.*, **81**, 4516.

Schuyten P. J. (1978). *New York Times, T.I. Sets Bubble Memory, Aug. 4, Section D, p.1.*.

Sharrock M. P. (1990). *IEEE Trans. Magn.*, **26**, 193.

Shliomis M. I. (1974). *Sov. Phys. Usp.*, **17**, 153.

Skjeltorp A. T. (1983). *Phys. Rev. Lett.*, **51**, 2306.

Slater J. C. (1930). *Phys. Rev.*, **36**, 57.

Slonczewski J. C. (1996). *J. Magn. Magn. Mater.*, **159**, L1.

Stiles M. D., McMichael R. D (1999). *Phys. Rev. B*, **59**, 372.

Stoner E. C., Wohlfarth E. P. (1948). *Phil. Trans. Roy. Soc. Lond. Series A, Mathematical and Physical Sciences*, **240**, 599.

Street R., McCormick P. G., Folks L. (1992). *J. Magn. Magn. Mater.*, **104-107**, 368.

Street R., Wolley J. C. (1949). *Proc. Phys. Soc. A*, **62**, 562.

Tao B. S., Yang H. X., Zuo Y. L., Devaux X., Lengaigne G., Hehn M., Lacour D., Andrieu S., Chshiev M., Hauet T., Montaigne F., Mangin S., Han X. F., Lu. Y. (2015). *Phys. Rev. Lett.*, **115**, 157204.

Tari A., Chantrell R. W., Charles S. W., Popplewell, J. (1979). *Physica B+C*, **97**, 57.

Tartaj P., del Puerto Morales M., Veintemillas-Verdaguer S., González-Carreño1 T., Serna C. J. (2003). *J. Phys. D: Appl. Phys.*, **36**, R182.

Thiesen B., Jordan A. (2008). *Int. J. Hyperthermia*, **24**, 467.

Tholence J. L., Tournier R. (1977). *Physica B+C*, **86-88**, 873.

Thomson S. M. (2008). *J. Phys. D: Appl. Phys.*, **41**, 093001.

Thompson D., Romankiw L., Mayadas A. (1975). *IEEE Trans. Magn.*, **11**, 1039.

Thomson W. (1857). *Proc. Roy. Soc.*, **8**, 546.

Tomeno I., Fuke H. N., Iwasaki H., Sahashi M. (1999). *J. Appl. Phys.*, **86**, 3853.

Tsunegi S., Taniguchi T., Lebrun R., Yakushiji K., Cros V., Grollier J., Fukushima A., Yuasa S., Kubota H. (2018). *Sci. Rep.*, **8**, 13475.

Tsunoda M., Imakita K., Naka M., Takahashi M., (2006). *J. Magn. Magn. Mater.*, **304**, 59.

Tsunoda M., Yoshitaki S., Ashizawa Y., Kim D. Y., Mitsumata C., Takahashi M. (2007). *Phys. Stat. Solidi. B*, **244**, 4470.

Umeki S., Saitoh S., Imaoka Y. (1974). *IEEE Trans. Magn.*, **10**, 655.

Uren S., Walker M., O'Grady K, Chantrell R. W. (1988). *IEEE Trans. Magn.*, **24**, 1808.

Vallejo-Fernandez G., Aley N. P., Fernandez-Outon L. E., O'Grady K. (2008c). *J. Appl. Phys.*, **104**, 033906.

Vallejo-Fernandez G., Chapman J. N. (2009). *Appl. Phys. Lett.*, **94**, 262508.

Vallejo-Fernandez G., Fernandez-Outon L. E., O'Grady K. (2007). *Appl. Phys. Lett.*, **91**, 212503.

Vallejo-Fernandez G., Fernandez-Outon L. E., O'Grady K. (2008b). *J. Phys. D: Appl. Phys.*, **41**, 112001.

Vallejo-Fernandez G., Frost W., O'Grady K. (2023). *In preparation.*

Vallejo-Fernandez G., Kaeswurm B., Fernandez-Outon L. E., O'Grady K. (2008a). *IEEE Trans. Magn.*, **44**, 2835.

Vallejo-Fernandez G., Kaeswurm B., O'Grady K. (2011). *J. Appl. Phys.*, **109**, 07D738.

Vallejo-Fernandez G., O'Grady K. (2013). *Appl. Phys. Lett.*, **103**, 142417.

Vallejo-Fernandez G., Whear O., Roca A. G., Hussain S., Timmis J., Patel V., O'Grady K. (2013). *J. Phys. D: Appl. Phys.*, **46**, 312001.

van der Heijden P. A. A., Maas T. F. M. M., de Wonge W. J. M., Kools J. C. S., Roozeboom F., van der Zaag P. (1998). *Appl. Phys. Lett.*, **72**, 492.

van Kooten M., de Haan S., Lodder J. C., Lyberatos A., Chantrell R. W.,Miles J. J. (1993). *J. Magn. Magn. Mater.*, **120**, 145.

Victora R., Liu Z., Huang P.-W., Ju G., Thiele J.-U. (2017). *US Patent* , Number 10249335.

Vopsaroiu M., Georgieva M., Grundy P. J., Vallejo-Fernandez G., Manzoor S., Thwaites M. J., O'Grady K. (2005). *J. Appl. Phys.*, **97**, 10N303.

Walker M., Mayo P. I., O'Grady K., Charles S. W., Chantrell R. W. (1993). *J. Phys.: Cond. Matter*, **5**, 2779.

Walker M. W., Lloyd-Evans E. (2015). *Methods Cell Biol.*, **126**, 21.

Wang Z. G., Nakamura Y. (1996). *IEEE Trans. Magn.*, **32**, 4022.

Watanabe K., Jinnai B., Fukami S., Sato H., Ohno H. (2018). *Nat. Comm.*, **9**, 663.

Weiler M., Sattel S., Giessen T., Jung K., Ehrhardt H., Veerasamy V. S., Robertson J. (1996). *Phys. Rev. B*, **53**, 1594.

Weiss P. (1906). *Comptes Rendus des Séances de l'Académie des Sciences*, **143**, 1136.

Weller D., Parker G., Mosendz O., Lyberatos A., Mitin D., Safonova N. Y., Albrecht M. (2016). *J. Vac. Sci. Technol. B*, **34**, 060801.

Wen Z., Sukegawa H., Furubayashi T., Koo J., Inomata K., Mitani S., Hadorn J. P., Ohkubo T., Hono K. (2014). *Adv. Mater.*, **26**, 6483.

Williams H. D., O'Grady K., el-Hilo M., Chantrell R. W. (1993). *J. Magn. Magn. Mater.*, **122**, 129.

Williams H. J., Bozorth R. M., Shockley W. (1949). *Phys. Rev.*, **75**, 155.

Williams M. L., Comstock R. L. (1971). *17th Ann. AIP Conf. Proc.*, **5**, 738.

Wohlfarth E. P. (1958). *J. Appl. Phys.*, **29**, 595.

Wohlfarth E. P. (1979). *Phys. Lett.*, **70A**, 489.

Wohlfarth E. P. (1983). *J. Phys. F: Met. Phys.*, **14**, L155.

Xiao G., Liou S. H., Levy A., Taylor J. N., Chien C. L. (1986). *Phys. Rev. B*, **34**, 7573.

Xi H. (2005). *J. Magn. Magn. Mater.*, **288**, 66.

Xue J., Victora R. H. (2000). *J. Appl. Phys.*, **87**, 6361.

Yosida K. (1957). *Phys. Rev.*, **106**, 893.

Yuasa S., Nagahama T., Fukushima A., Suzuki Y., Ando K. (2004). *Nat. Mater.*, **3**, 868.

Zhu J.-G., Bertram H. N. (1989). *J. Appl. Phys.*, **66**, 1291.

Zhu J.-G., Bertram H. N. (1991). *J. Appl. Phys.*, **69**, 4709.

Zhu J.-G., Zhu X., Tang Y. (2006). *IEEE Trans. Magn.*, **44**, 125.

Index